AN ENGINEER'S VIEW OF HUMAN ERROR

By TREVOR A. KLETZ
(A Professor of Chemical Engineering, University of Technology, Loughborough)

The theme of this book:

Try to change situations, not people.

THE INSTITUTION OF CHEMICAL ENGINEERS
GEORGE E. DAVIS BUILDING
165-171 RAILWAY TERRACE
RUGBY
WARWICKSHIRE CV21 3HQ
ENGLAND

(i)

FOREWORD

In this book I set down some views on human error as a cause of accidents and illustrate them by describing a number of accidents that have occurred, mainly in the oil and chemical industries. Though the book is particularly addressed to those who work in the process industries, I hope that it will interest all those who design, construct, operate and maintain plant of all sorts, that it may suggest to some readers new ways of looking at accident prevention and that it will reinforce the views of those readers who already try to change situations rather than people. It is intended for practising engineers, especially chemical engineers, not for experts in human factors and ergonomics, and I hope it will be of particular interest to students.

Although 'human factors' are not specifically mentioned in The Institution of Chemical Engineers' model syllabus for chemical engineering undergraduate courses, they are an essential part of "Management for safety", which is included. Many of the accidents I describe could be used as the raw material for discussions on the causes of accidents and the action needed to prevent them happening again, using the methods described in the Institution's Hazard Workshop Modules. The Institution have produced a Module – a set of slides and notes – to accompany this book. A few of the incidents are already included in the existing modules.

Some readers may wonder why I have added to the existing literature on human error. When so much has been written, is there need for more?

I felt that much of the existing literature is too theoretical for the audience I have in mind, or devoted to particular aspects of the problem. I felt there was a need for a little book which would suggest to engineers how they might approach the problem of human error and do so by describing accidents which at first sight seem to be the result of human error. My approach is therefore pragmatic rather than theoretical.

I do not claim any great originality for the ideas in this book – many of them are to be found in other books such as *Man-Machine Engineering* by A Chapanis (Tavistock, 1965) – but while that book is primarily concerned with mechanical and control equipment, this one is concerned mainly with process equipment.

Not all readers may agree with the actions I propose. If you do not, may I suggest that you decide what action you think should be taken. Please do not ignore the accidents. They happened and will happen again, unless action is taken to prevent them happening.

Thanks are due to the many friends and colleagues, past and present, who suggested ideas for this book or commented on the draft – without their contributions I could not have produced this book – and to Mr S Coulson who prepared many of the illustrations. Thanks are also due to the many companies who allowed me to describe their mistakes and to the Science and Engineering Research Council for financial support.

TREVOR A KLETZ

CONTENTS

CHAPTER 1

INTRODUCTION

*"Man is a creature made at the end of the week . . .
when God was tired"* — Mark Twain

1.1 Accept men as we find them

The theme of this book is that it is difficult for engineers to change human nature and therefore, instead of trying to persuade people not to make mistakes, we should accept people as we find them and try to remove opportunities for error by changing the work situation, that is, the plant or equipment design or the method of working. Alternatively, we can mitigate the consequences of error or provide opportunities for recovery. A second objective of the book is to remind engineers of some of the quirks of human nature so that they can better allow for them in design.

The method used in the book is to describe accidents which at first sight were due to human error and then discuss the most effective ways of preventing them happening again. The accidents occurred mainly, though not entirely, in the oil and chemical industries, but nevertheless should interest all engineers, not just chemical engineers, and indeed all those who work in design or production.

Browsing through old ICI files I came across a report dating from the late 1920's in which one of the Company's first safety officers announced a new discovery: after reading many accident reports he had realised that most accidents are due to human failing. The remedy was obvious. We must persuade people to take more care.

Since then people have been exhorted to do just this, and this policy has been supported by tables of accident statistics from many companies which show that over 50%, sometimes as many as 90%, of industrial accidents are due to human failing. This is comforting for managers. It implies that there is little or nothing they can do to stop most accidents.

In the early 1960's (when I was a manager, not a safety adviser) I looked through a bunch of accident reports and realised that most of them could be prevented by better management — sometimes by better design or method of working, sometimes by better training or instructions, sometimes by better enforcement of the instructions. Together these may be called changing the work situation. There was, of course, an element of human failing in the accidents. They would not have occurred if someone had not forgotten to close a valve, looked where he was going, not taken a short-cut. But what chance do we have, without management action of some sort, of persuading people not to do these things?

To say that accidents are due to human failing is not so much untrue as unhelpful. It does not lead to any constructive action. All we can do is tell someone to be more careful. In contrast if we say that an accident can be prevented by better design, or by better training or instructions, or by better auditing or inspection, we can take action that may prevent a recurrence.

I would rather say that an accident can be prevented by better design, better instructions, etc, than say it was caused by bad design, instructions, etc. Cause implies blame and we become defensive. We do not like to admit that we did something badly, but we are willing to admit that we could do it better.

I do not say that it is impossible to change people. Those more qualified than engineers to do so — teachers, clergymen, social workers, psychologists — will no doubt continue to try and we wish them success. But the results achieved in the last few thousand years suggest that their results will be neither rapid nor spectacular and where experts achieve so little, engineers are likely to achieve less. Let us therefore accept that people are the one component of the systems we design that we cannot redesign or modify. We can design better pumps, compressors, distillation columns, etc, but we are left with Mark I man (and woman).

1.2   Meccano or dolls?

Let me emphasise that when I suggest changing the work situation, I am not simply saying change the hardware. Sometimes we have to change the software — the method of working, training, instructions, etc. Safety by design should always be our aim, but sometimes redesign is impossible, or too expensive, and we have to modify procedures. In over half the accidents that occur there is no reasonably practical way of preventing a repetition by a change in design and we have to change the software.

We can change people's performance by better training, instructions, supervision, etc. What we cannot change is the propensity of men to have a moment's aberration and make a simple mistake or to carry out tasks beyond their physical or mental powers. These errors we have to accept — or remove the opportunities for them to occur.

At present, most engineers are men and as boys most of us played with Meccano rather than dolls. We were interested in machines and the way they work, otherwise we would not be engineers. Most of us are very happy to devise hardware solutions. We are less happy when it comes to software solutions, to devising new training programmes or methods, writing instructions, persuading people to follow them, checking up to see that they are being followed and so on. However, these solutions are just as important as the hardware ones, as we shall see, and require as much of our effort and attention.

One reason we are less happy with software solutions is that continual effort — what I have called grey hairs[1] — is needed to prevent them disappearing. If we can remove a hazard by modifying the hardware or installing extra hardware, we may have to fight for the money, but once we get it and the equipment is modified or installed it is unlikely to disappear.

In contrast, if we remove a hazard by modifying a procedure or introducing extra training, we may have less difficulty getting approval, but the new procedure or training programme may vanish without trace in a few months once we lose interest. Procedures are subject to a form of corrosion more rapid and thorough than that which affects the steelwork. Procedures lapse, trainers leave and are not replaced. A continuous management effort — grey hairs — is need to maintain our systems. No wonder we prefer safety by design whenever it is possible and economic; unfortunately, it is not always possible and economic.

Furthermore, when we do go for safety by design, the new equipment may have be tested and maintained. It is easy to install new trip systems — all you have to do is persuade someone to give you the money. You will get more grey hairs seeing that they are tested and maintained and that people are trained to use them properly and do not try to disarm them.

1.3   Types of human error

In the following pages I shall discuss different sorts of human error, describe accidents caused by them and suggest ways of preventing similar accidents in the future. The sorts of errors I shall consider are:

— Errors due to slips or aberrations (Chapter 2)

— Errors that could be prevented by better training or instructions (Chapter 3)

— Errors due to a lack of physical or mental ability (Chapter 4)

— Errors due to a lack of motivation, in particular, errors preventable by better supervision (Chapter 5).

In Chapter 6 we shall look at some of the attempts that have been made to quantify the probability of human error. However, these methods apply only to the first sort of error. We can estimate — roughly — the probability that someone will have a moment's aberration and forget to open a valve, or open the wrong valve, but we cannot estimate the probability that he will make a mistake because the training or instructions are poor, because he lacks the necessary physical or mental ability, or because he has a 'couldn't care less' attitude. Each of these factors can contribute from 0 to 100% to the probability of failure. People often assume that these errors are eliminated by selection, training, instructions and monitoring, but this is not always true.

The allocation of accidents to these categories is to some extent arbitrary — more than one factor is at work in many incidents — and for this reason in Chapters 7-10 we shall look at some further accidents due to human failing, but classified somewhat differently. We shall consider:

— Accidents that could be prevented by better design

— Accidents that could be prevented by better construction

— Accidents that could be prevented by better maintenance

— Accidents that could be prevented by better methods of operation.

Finally we shall look at legal views and at the question of personal responsibility. If we try to prevent errors by changing the work situation, does this mean that people who make errors are entitled to say "It is your fault for putting me in a situation where I was able to make a mistake"?

1.4   A simple example

The various types of human error may be illustrated by considering a simple everyday error: forgetting to push in the choke on a car when the engine has warmed up. There are several possible reasons for the error:

It may be due to forgetfulness. This is the most likely reason and similar errors are discussed in Chapter 2.

With an inexperienced driver the error may be due to a lack of training or instruction; he may not have been told what to do or when to do it. Similar errors are discussed in Chapter 3.

The error may be due to a lack of physical or mental ability — unlikely in this case. Such errors are discussed in Chapter 4.

4

The error may be due to the fact that the driver cannot be bothered — the hired car syndrome. Similar errors are discussed in Chapter 5.

Chapter 6 discusses the probability of the first sort of error.

If the error is due to lack of training or instruction then we can provide better training or instruction. There is not much we can do in the other cases except change the work situation, that is, provide a warning light or alarm or an automatic choke. The latter adds to the cost and provides something else to go wrong and it might be better to accept the occasional error. Examples of situations in which automatic equipment is not necessarily better than an operator are given in Chapter 6.

1.5 A story

The following story illustrates the theme of this book.

A man went into a tailor's shop for a ready-made suit. He tried on most of the stock without finding one that fitted him. Finally, in exasperation, the tailor said, "I'm sorry, sir, I can't fit you; you're the wrong shape".

Should we as engineers expect men to change their (physical or mental) shapes so that they fit into the plants and procedures we have designed or should we design plants and procedures to fit men?

1.6 Research on human error

The Third Report of the Advisory Committee on Major Hazards[2] recommends research on human reliability and makes some suggestions. While there is undoubtedly much we would like to know, lack of knowledge is not the main problem. The main problem is that we do not use the knowledge we have. Accidents, with a few exceptions, are not caused by lack of knowledge, but by a failure to use the knowledge that is available. This book is an attempt, in a small way, to contribute towards accident reduction by reminding readers of facts they probably know well and re-inforcing them by describing accidents which have occurred, in part, as a result of ignoring these facts.

References to Chapter 1

1.   T A Kletz, *Plant/Operations Progress*, Vol 3, No 4, 1984, p 210.

2.   Advisory Committee on Major Hazards, *Third Report, The Control of Major Hazards*, HMSO, 1984, Appendix 13.

CHAPTER 2

ACCIDENTS CAUSED BY SIMPLE SLIPS

"I haven't got a memory, only a forgettory" —
small boy quoted on *Tuesday Call*, BBC
Radio 4, 6 December 1977

2.1 Introduction

In this Chapter I describe some accidents which occurred because someone forgot to carry out a simple, routine step such as closing or opening a valve, or carried it out wrongly, that is, he closed or opened the wrong valve. He knew what to do, had done it many times before, was capable of doing it and intended to do it and do it correctly but he had a moment's aberration.

Such slips are similar to the slips of everyday life and cannot be prevented by exhortation, punishment or further training. We must either accept an occasional mistake — probabilities are discussed in Chapter 6 — or remove the opportunity for error by a change in the work situation, that is, by changing the design or method of working. Alternatively, we can, in some cases, provide opportunities for people to observe and correct their errors or we can provide protection against the consequences of error.

Note that errors of the type discussed in this Chapter occur not in spite of the fact that the man who makes it is well-trained but *because* he is well-trained. Routine tasks are given to the lower levels of the brain and are not continuously monitored by the conscious mind. We could never get through the day if everything we did required our full attention, so we put ourselves on auto-pilot. When the normal pattern or programme of action is interrupted for any reason, errors are liable to occur.

Those interested in the psychological mechanisms by which errors occur should read J Reason and C Mycielska's book on everyday errors.[1] The psychology is the same. Here, as engineers, we are not concerned with the reasons why errors occur, but with the fact that they do occur and that we can do little to prevent them. We should therefore accept them and design accordingly.

People who have made an absent-minded error are often told to keep their mind on the job. It is understandable that people will say this, but it is not very helpful. No one deliberately lets their mind wander, but it is inevitable that we all do so from time to time.

Can we reduce errors by selecting people who are less error-prone? Obviously if anyone makes a vast number of errors — far more than an ordinary person would make — there is something abnormal about his work situation or he is not suitable for the job. But that is as far as it is practicable to go. If we could identify people who are slightly more forgetful than the average person — and it is doubtful if we could — and weed them out, we might find that we have weeded out the introverts and left the extroverts and those are the people who cause accidents for another reason: they are more inclined to take chances. We return to this subject later (Section 4.3).

2.2 Forgetting to open or close a valve

2.2.1 Example 1 — Opening equipment which has been under pressure

A suspended catalyst was removed from a process stream in a pressure filter (Figure 2.1). When a batch had been filtered, the inlet valve was closed and the liquid in the filter blown out with steam. The steam supply was then isolated, the pressure blown off through the vent and the fall in pressure observed on a pressure gauge. The operator then opened the filter for cleaning. The filter door was held closed by eight radial bars which fitted into U-bolts on the filter body. To withdraw the radial bars from the U-bolts and open the door the operator had to turn a large wheel, fixed to the door. The door, with filter leaves attached, could then be withdrawn (Figure 2.2).

One day an operator, a conscientious man of great experience, started to open the door before blowing off the pressure. He was standing in front of it and was crushed between the door and part of the structure and was killed instantly.

This accident occurred some years ago and at the time it seemed reasonable to say that the accident was due to an error by the operator. It showed the need, it was said, for other operators to remain alert and to follow the operating instructions exactly. Only minor changes were made to the design.

However, we now see that in the situation described it is inevitable, sooner or later, that an operator will forget that he has not opened the vent valve and will try to open the filter while it is still under pressure. The accident was the result of the work situation, and we would recommend the following changes in the design:

(1) Whenever someone has to open up equipment which has been under pressure, using quick release devices, (a) Interlocks should be fitted so that the vessel cannot be opened until the source of pressure is isolated and the vent valve opened (one way of doing this would be to arrange for the handles of ball valves on the steam and vent lines to project over the door handle when the steam valve is open and the vent valve closed) and (b) The design of the door or cover should be such that it can be opened about ¼ inch (6 mm) while still capable of carrying the full pressure and a

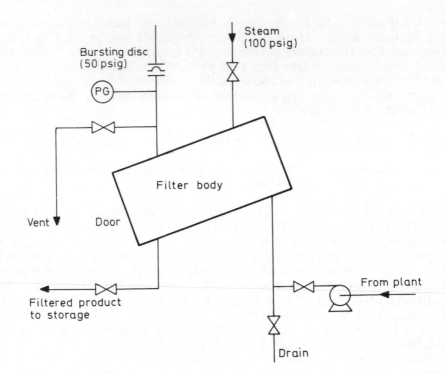

Figure 2.1   Filter.

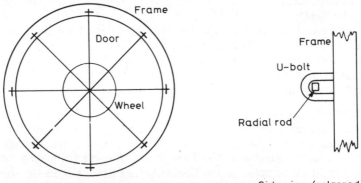

Figure 2.2   Filter door and fastening.

separate operation should be required to release the door fully. If the cover is released while the vessel is under pressure, this is immediately apparent and the pressure can be allowed to blow off through the gap or the door can be resealed. [2,3]

(2) The pressure gauge and vent valve should be located near the door so that they are clearly visible to the operator when he is about to open the door. They were located on the floor above.

(3) The handle on the door should be modified so that it can be operated without the operator having to stand in front of the door.

Recommendations (2) and (3) were made and carried out at the time but not (1), the most important.

The accident occurred at the end of the night shift, an hour before the operator was due to start his annual holiday. His mind may not have been fully on the job; he may have been thinking of his holiday. Who can blame him? It is not very helpful to say that the accident was due to human failing. We all have moments when for one reason or another our minds are not on the job. This is inevitable. We should list as the causes of an accident only those we can do something about. In this case the accident could have been prevented by better design of the equipment. [4]

Many similar accidents have occurred when operators have had to open up equipment which has been under pressure. In contrast, every day, in every factory, equipment which has been under pressure is opened up for repair but this is normally done under a permit-to-work system. One man prepares the equipment and issues a permit to another man who opens it up, normally by carefully slackening bolts in case there is any pressure left inside. Safety is obtained by following procedures: The involvement of two men and the issue of a permit provides an opportunity to check that everything necessary has been done. Accidents are liable to happen when the same man prepares the equipment and opens it up and in these cases we have to look for safety by design. This is now required in the UK by the Health and Safety Executive. [2,3]

One design engineer, finding it difficult to install the devices recommended in (1) above, said that it was "reasonable to rely on the operator". He would not have said it was reasonable to rely on the operator if a tonne weight had to be lifted; he would have installed mechanical aids. Similarly if memory tasks are too difficult we should install mechanical (or procedural) aids.

Of course, we can rely on the operator to open a valve 99 times out of 100, perhaps more; perhaps less if stress and distraction are high (see Chapter 6), but one failure in a hundred or even in a thousand, is far too high when we are dealing with an operation which is carried out every day and where failure can have serious results.

Many design engineers accept the arguments of this book in principle but when safety by design becomes difficult they relapse into saying, "We shall have to rely on the operator". They should first ask what failure rate is likely and whether that rate is acceptable. (See Chapter 6.)

Would a check list reduce the chance that someone will forget to open a valve?

Check lists are useful when performing an unfamiliar task, for example, a plant start-up or shut-down which occurs only once per year. It is unrealistic to expect people to use them when carrying out a task which is carried out every day or every few days. The operator knows exactly what to do and sees no need for a check list. If the manager insists that one is used, and that each step is ticked as it is completed, the list will be completed at the end of the shift.

We all forget occasionally to push in the chokes on our cars when the engines get hot but we would not agree to complete a check list every time we start our cars.

It is true that aircraft pilots go through a check list at every take-off, but the number of checks to be made is large, the consequences of failure serious and pilots are more highly trained and educated than process operators.

2.2.2 Example 2 — Emptying a vessel

Figure 2.3 shows the boiler of a batch distillation column. After a batch was complete the residue was discharged to a residue tank through a drain valve which was operated electrically from the control room. To reduce the chance that the valve might be opened at the wrong time, a key was required to operate the valve and indicator lights on the panel showed whether it was open or shut.

One day the operator, while charging the still, noticed that the level was falling instead of rising and then realised that he had forgotten to close the drain valve after emptying the previous batch. A quantity of feed passed to the residue tank where it reacted violently with other residues.

The operator pointed out that the key was small and easily overlooked and that the indicator lights were not easily visible when the sun was shining. As an immediate measure a large metal tag was fitted to the key. Later the drain and feed valves were interlocked so that only one of them could be open at a time.

A similar incident occurred on the reactor shown in Figure 2.4. When the reaction was complete the pressure fell and the product was discharged into the product tank. To prevent the discharge valve being opened at the wrong time, it was interlocked with the pressure in the reactor so that it could not be opened until the pressure had fallen below a gauge pressure of 0.3 bar.

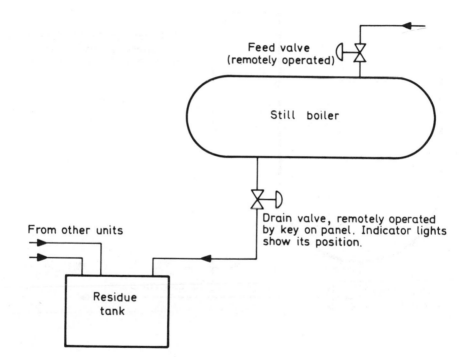

Figure 2.3 Premature discharge of the boiler caused a reaction to occur in the — residue tank.

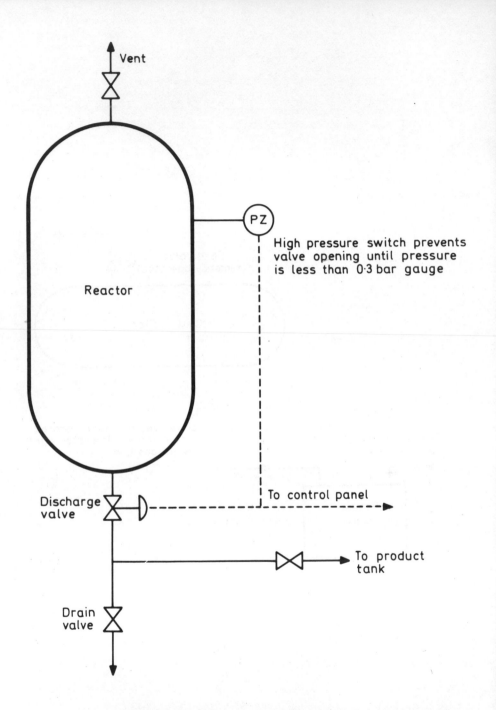

Vent

PZ

High pressure switch prevents
valve opening until pressure
is less than 0·3 bar gauge

Reactor

To control panel

Discharge
valve

To product
tank

Drain
valve

Figure 2.4 Arrangement of valves on a batch reactor. The trip was left in the
open position so that when the pressure fell the discharge valve opened.

12

A batch failed to react and it was decided to vent off the gas. When the gauge pressure fell below 0.3 bar, the discharge valve opened automatically and, as the drain valve was also open, the contents of the reactor were discharged into the working area. Fortunately, though they were flammable, they did not catch fire. The remote actuator on the discharge valve had been left in the open position and as soon as the pressure in the reactor fell the interlock allowed the valve to open.

The drain valve was provided for use after the reactor was washed out with water between batches.

Again, it is too simplistic to say that the accident was the result of an error by the operator who left the actuator on the discharge valve in the open position. The accident could have been prevented by a better designed protective system.

Could the weakness in the design of the protective system have been foreseen by a hazard and operability study?

2.2.3   Example 3 – Emptying a pump

A pump had to be drained when not in use and the drain valve left open as the liquid inside it gave off gas on standing and the pressure would damage the seal. Before starting up the pump, the drain valve had to be closed.

One day a young operator, in a hurry to start up the spare pump and prevent interruption to production, forgot to close the drain valve although he knew that he should, had done it before and an instruction to do so was included in the plant operating instructions. The liquid came out of the open drain valve and burnt him chemically on the leg. The accident was said, in the report, to be due to human failing and the operator was told to be more careful.

However, changes were made to make the accident less likely and protect the operator from the consequences of error. A notice was put up near the pump to remind the operator that the drain valve was normally open and the drain line was moved so that the drainings were still visible but less likely to splash the operator.

2.3   Operating the wrong valve

2.3.1   Trip and alarm testing

A plant was fitted with a low pressure alarm and an independent low pressure trip, arranged as shown in Figure 2.5. There was no label on the alarm but there was a small one near the trip.

An instrument artificer was asked to carry out the routine test of the alarm. The procedure, well-known to him, was to isolate the alarm from the plant, open the vent to blow off the pressure and note the pressure at which the alarm operated.

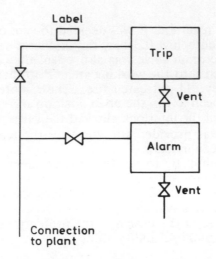

Figure 2.5 Arrangement of trip and alarm connections.

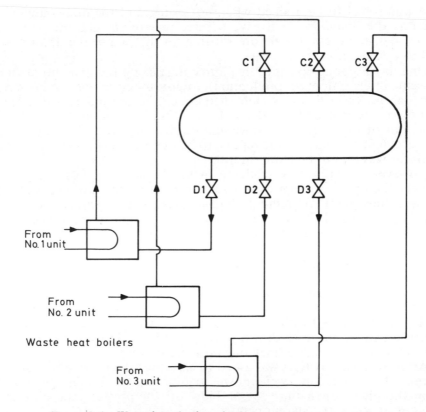

Figure 2.6 Waste heat boilers sharing a common steam drum.

By mistake he isolated and vented the trip. When he opened the vent valve, the pressure in the trip fell and the plant was automatically shut down. It took 36 hours to get it back to normal.

It would have been little use telling the artificer to be more careful. To reduce the chance of a further mistake we should:

(a)   Provide better labels

(b)   Put the trip and alarm further apart

(c)   Possibly paint the trip and alarm different colours.

Though not relevant to this incident, note that the trip and alarm should be connected to the plant by separate impulse lines to reduce the chance of a common mode failure: choking of the common impulse line.

2.3.2   Isolation of equipment for maintenance

To save cost three waste heat boilers shared a common steam drum. Each boiler had to be taken off line from time to time for cleaning (Figure 2.6).   On two occasions the wrong valve was closed (D3 instead of D2) and an on-line boiler was starved of water and over-heated.   On the first occasion the damage was serious.   High temperature alarms were then installed on the boilers.   On the second occasion they prevented serious damage but some tubes still had to be changed.   A series of interlocks were then installed so that a unit has to be shut down before a key can be removed; this key is needed to isolate the corresponding valves on the steam drum.

The chance of an error was increased by the lack of labelling and the arrangement of the valves — D3 was below C2.

A better design, used on later plants, is to have a separate steam drum for each waste heat boiler (or group of boilers if several can be taken off line together).   There is then no need for valves between the boiler and the steam drum.   This is more expensive but simpler and free from opportunities for error.

2.4   Pressing the wrong button

2.4.1   Example 1 — Beverage machines

Many beverage vending machines are fitted with a panel such as that shown in Figure 2.7.   I found that on about one occasion in 50 when I used these machines I pressed the wrong button and got the wrong drink.   Obviously the consequences were trivial and not worth worrying about, but suppose a similar panel was used to charge a batch reactor or fill a container with product for sale: Pressing the wrong button might result in a runaway or unwanted reaction or a customer complaint.

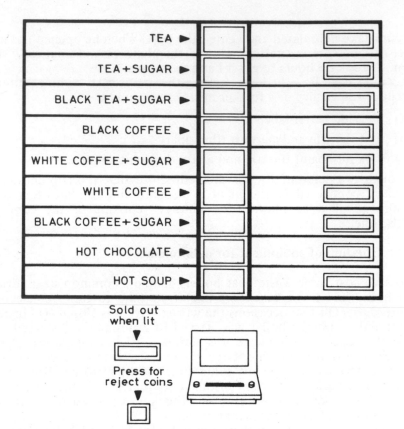

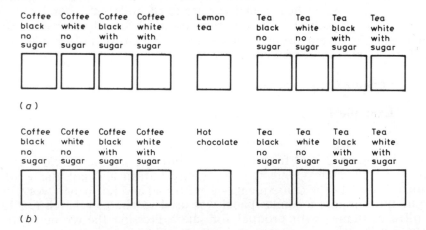

Figure 2.7 Panel of beverage vending machine.

Coffee black no sugar	Coffee white no sugar	Coffee black with sugar	Coffee white with sugar	Lemon tea	Tea black no sugar	Tea white no sugar	Tea black with sugar	Tea white with sugar
□	□	□	□	□	□	□	□	□

(*a*)

Coffee black no sugar	Coffee white no sugar	Coffee black with sugar	Coffee white with sugar	Hot chocolate	Tea black no sugar	Tea white no sugar	Tea black with sugar	Tea white with sugar
□	□	□	□	□	□	□	□	□

(*b*)

Figure 2.8 Panels of two similar beverage vending machines.

16

It is therefore worth studying the factors that influence the probability of error and ways of reducing it.

This example of human error is of interest because many of the uncertainties that are present in other examples do not apply. The errors were not due (I hope) to lack of training or instructions, or to lack of physical or mental ability. They were certainly not due to lack of motivation because when I got the wrong drink I had to drink it (being too mean or short of change to throw it away and try again). I knew what to do, was able to do it and wanted to do it, but nevertheless made an occasional mistake.

My error rate was increased by a certain amount of stress and distraction. (The machines are in the corridor.) It is shown in Section 6.5 that in a situation free from stress and distraction the error rate would probably be about 3 in 1000.

If we wished to reduce the error rate we could:

— Place the machine in a place where there is less distraction; stress is harder to remove.

— Redesign the panel. We could put the buttons further apart.

— Better still, we could separate the choice of drink (tea, coffee, chocolate, soup) from the choice of milk (yes or no) and sugar (yes or no).

If this did not give an acceptable error rate we might have to consider attaching a microprocessor which could be told the name of the product or customer and would then allow only certain combinations of constituents. Alternatively it might display the instructions on a screen and the operator would then have to confirm that they were correct.

I moved to a new building where the beverage machines were of a different type, shown in Figure 2.8(a). I started to drink lemon tea — obtained by pressing the isolated central button — and I felt sure that I would make no more mistakes. Nevertheless, after a few weeks, I did get the wrong drink. I used a machine in a different part of the building and when I pressed the centre button I got hot chocolate. The panel was arranged as shown in Figure 2.8(b).

The labelling was quite clear, but I was so used to pressing the centre button that I did not pause to read the label.

A situation like this sets a trap for the operator as surely as a hole outside the control room door. Obviously it does not matter if I get the wrong drink but a similar mistake on a plant could be serious. If there are two panels in the control room and they are slightly different, we should not blame the operator if he makes a mistake. The panels should be identical or entirely different. If they have to be slightly different — because of a different function — then a striking notice or change of colour is needed to draw attention to the difference.

Two identical units shared a common control room. The two panels were arranged as mirror images of each other. This led to numerous mistakes.

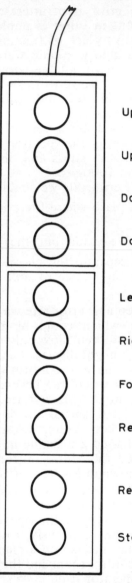

Figure 2.9   Control unit for a travelling overhead crane.

There is an extensive literature on the ergonomics of control room layout.

2.4.2   Example 2 – Overhead cranes

Figure 2.9 shows the buttons on the control unit for an overhead crane. Operators become very skilled in their use and are able to press two or three buttons at a time so that they seem to be playing the controls like a concertina. Nevertheless occasional errors occur. However, in this case the operator sees the load move the wrong way and can usually reverse the movement before there is an accident. Nevertheless we must expect an occasional 'bump'.

2.4.3   Example 3 – Charging a reactor

A similar mistake to the one with the coffee machine caused a serious fire in which several men were killed.

Two reactors, Nos 4 and 6, on a batch plant were shut down for maintenance. The work on No 4 was completed and the foreman asked an operator to open the feed valve to No 4. The valve was electrically operated and the operator went to the panel, shown in Figure 2.10, but by mistake pressed the button controlling the inlet valve to No 6 reactor . This reactor was still under repair. The valve opened; flammable gas came out and caught fire. The company concerned said, "What can we do to prevent men making mistakes like this?" The answer is that we cannot prevent men making such mistakes, though we can make them less likely by putting the buttons further apart, providing better labelling and so on, but mistakes will still happen, particularly if the operator is under stress (Figure 12.7).

We should never tolerate a situation in which such a simple slip has such serious consequences. The valves on reactors under maintenance should have been defused and locked off and the inlet and exit lines should been slip-plated.

The operator was *not* to blame for the accident. He made the sort of mistake that everyone makes from time to time. The accident could have been prevented by a better method of working, by better management.

2.4.4   Example 4 – Shutting down equipment

A row of seven furnaces was arranged as shown in Figure 2.11(a).

The buttons for isolating the fuel to the furnaces were arranged as shown in Figure 2.11(b).

An operator was asked to close the fuel valve on No 5 furnace. By mistake he pressed the wrong button and isolated the fuel to A furnace.

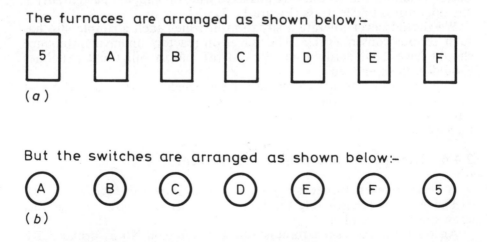

Figure 2.10 Arrangement of operating buttons for the inlet and exit valves on a group of batch reactors.

The furnaces are arranged as shown below:-

| 5 | A | B | C | D | E | F |

(a)

But the switches are arranged as shown below:-

(A) (B) (C) (D) (E) (F) (5)

(b)

Figure 2.11 Arrangement of furnaces and control panel.

He realised that he had to isolate the fuel to the furnace on the extreme left — so, without thinking, he went to the button on the extreme left.

2.5   Failures to notice

The failures discussed so far have been failures to carry out an action — such as forgetting to close a valve — or carrying it out incorrectly — closing the wrong valve.

Accidents can occur because someone fails to notice the signal for an action. We should not tell people to "Wake up" or "Pay more attention", but should make the signal more prominent.

For example, a company fitted small discs to every ladder, giving the date on which they were due for inspection. They were fitted to the top rungs.

Someone pointed out that they were more likely to be seen if they were fitted to the sixth rung from the bottom!

2.6   Errors in calculations

A batch reaction took place in the presence of an inorganic salt which acted as a buffer to control the pH. If it was not controlled, a violent exothermic side-reaction occurred. Each batch was 'tailor-made' for the particular purpose for which the product was required and the weights of the raw materials required were calculated from their compositions and the product specification.

As the result of an error in calculation, only 60% of the buffer required was added. There was a runaway reaction and the reactor exploded. Following the explosion there was a review of the adequacy of the protective system and many additional trips were added. On the modified plant, loss of agitation, high temperature or low pH resulted in the reaction being automatically aborted by addition of water. A level switch in the water tank prevented operation of the charge pump if the level was low.

In addition, the amount of buffer added was increased to twice that theoretically necessary. However, errors in calculation were still possible, as well as errors in the quantities of reactants added, and so there was a need for the protective instrumentation.

An error by a designer resulted in the supports for a small tank being too light. When the tank was filled with water for a construction pressure test, it fell to the ground, unfortunately fracturing an oil line and causing a fire which killed a man.

In another case an experienced foreman said that the supports for a new pipebridge were too far apart. The manager (myself) said, "What you mean is that they are further apart than in the past. New methods of calculation", I suggested, "resulted in a cheaper design".

After a flanged joint on a pipeline on the bridge had leaked it was found that there had been an error in calculation and an extra support had to be added.

Despite these two incidents, calculation errors by designers are comparatively rare and I do not suggest any changes to normal design procedures. But anything that looks odd after construction should be checked. *What does not look right, may not be right.*

2.7 Other industries

By considering accidents due to human error in other industries we can check the validity of our conclusion — that occasional errors are inevitable and that we must either accept them or change the work situation, but that it is little use exhorting people to take care. Because we are not involved, and do not have to rethink our designs or modify our plants, we may see the conclusion more clearly.

2.7.1 Railways

Railways have been in existence for a long time, railway accidents are well-documented in the official reports by the Railway Inspectorate and in a number of books [5-8] and they provide examples illustrating all the principles of accident investigation, not just those concerned with human error. Those interested in industrial safety will find that the study of railway accidents is an enjoyable way of increasing their knowledge of accident prevention.

2.7.1.1 Signalmen's errors

A signalman's error led to Britain's worst railway accident, at Quintinshill just north of the Scottish border on the London — Glasgow line, in 1915, when 226 people were killed, most of them soldiers.[9]

Figure 2.12 shows the layout of the railway lines. Lines to London are called up lines; lines from London are called down lines.

The two loop lines were occupied by goods trains and so a slow north-bound passenger train was backed on to the up-line in order to let a sleeping car express come past. The signalman, who had just come on duty, had had a lift on the slow train and had jumped off the footplate as it was backing on to the up-line. He could see the slow train through the signalbox window. Nevertheless, he completely forgot about it and accepted a south-bound troop train which ran into the slow train. A minute or so later the north bound express train ran into the wreckage. The wooden coaches of the troop train caught fire and many of those who survived the first impact were burned to death.

The accident occurred because *the signalman forgot that there was a train on the up line,* though he could see it from his window and had just got off it, and accepted another train.

A contributory cause was the failure of the signalman who had just gone off duty to inform the signalman in the next signalbox that the line was blocked and to put a reminder collar on the signal lever.

One signalman had a lapse of memory — and obviously it was not deliberate.

The other signalman was taking short cuts — omitting to carry out jobs which he may have regarded as unnecessary.

What should we do?

As in the other incidents discussed, there are three ways of preventing similar incidents happening again:

    i)   Change the hardware,

    ii)   Persuade the operators to be more careful,

    iii)   Accept the occasional accident (perhaps taking action to minimise the consequences).

i)   Changing the hardware was in this case, possible, but expensive. The presence of a train on a line can complete a "track circuit" which prevents the signal being cleared. At the time, track circuiting was just coming into operation and the Inspector who conducted the official enquiry wrote that Quintinshill, because of its simple layout, would be one of the last places where track circuiting would be introduced. It was not, in fact, installed there until the electrification of the London — Glasgow line many years later; many lines are still not track-circuited.

ii)   Both signalmen were sent to prison — it was war-time and soliders had been killed.

It is doubtful if prison, or the threat of it, will prevent anyone forgetting that there is a train outside their signalbox, especially if they have just got off it.

Prison, or the threat of it, might prevent people taking short cuts, but a better way is management supervision. Did anyone check that the rules were followed? It would be surprising if the accident occurred on the first occasion on which a collar had been left off or another signalman not informed. (See Chapter 5.)

iii)   In practice, since prison sentences were probably ineffective, society accepted that sooner or later other similar accidents will occur. They have done, though fortunately with much less serious consequences.

Sometimes accepting an occasional accident is the right solution, though we do no like to admit it. So we tell people to be more careful if the accident is trivial, punish them if the accident is serious and pretend we have done something to prevent the

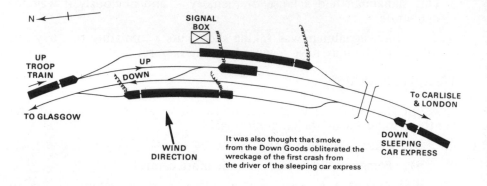

Figure 2.12 Layout of lines at Quintinshill, the scene of Britain's worst railway accident.

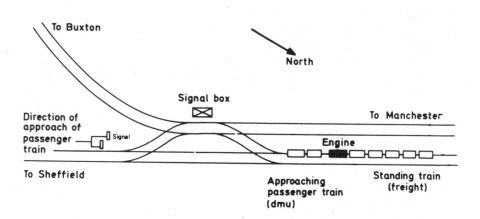

Figure 2.13 Track layout at Chinley North Junction. The signalman forgot that the freight train was standing on the track and allowed the passenger train to approach.

accident recurring. In fact, the accidents arise out of the work situation and, if we cannot accept an occasional accident, we should change the work situation.

Although a similar accident could happen today on the many miles of British Rail track which are not track-circuited, the consequences would be less serious as modern all-steel coaches and modern couplings withstand accidents much better than those in use in 1915.

A similar accident to Quintinshill occurred recently on British Railways, though the consequences were less serious. A signalman forgot there was a freight train standing on the wrong line and accepted another train (Figure 2.13). The track-circuiting prevented him releasing the signal so he assumed there was a fault in the track-circuiting and displayed a green hand signal to the driver of the oncoming train. He did not even check the illuminated diagram in his signalbox which would have shown the standing train. [10]

As at Quintinshill, there was an irregularity in the signalman's procedures. He should have gone down to the track to give the green hand signal, not displayed it from the signalbox. Had he done so, he might have seen the standing train.

This incident shows how difficult it is to design protective equipment which is proof against all human errors. If signalmen are not permitted to use green hand lamps, what do they do when track-circuiting and signals go out of order? The incident is also an example of a mind-set, discussed later (Section 4.4). Once we have come to a conclusion we close our minds to further evidence and do not carry out the simplest checks.

One change since 1915 is that there was no suggestion that the signalman should be punished. Instead the official report wonders if his domestic worries made an error more likely.

2.7.1.2 Drivers' errors

Many accidents have occurred because drivers passed signals at danger. At Aisgill in the Pennines in 1913, 14 people were killed and the driver, who survived, was imprisoned. At Harrow in 1952, 112 people, including the driver, were killed. At Moorgate in 1975, 42 people, including the driver, were killed. The driver is the person most at risk and many other drivers have been killed in this way, so they have no incentive to break the rules, but accidents still occur. Many are described in recent official reports.

Davis [11] has analysed a number of cases in detail and has shown that while a few of the drivers were clearly unsuited to the job, the majority were perfectly normal men with many years' experience who had a moment's aberration.

As with signalmen, we should accept an occasional error (the probability is discussed in Section 6.8) — this may be acceptable on little-used branch lines — or install a form of automatic train control. On the British Railways system a hooter sounds when a driver approaches a distant signal at danger (yellow) and if he does not acknowledge the signal, the brakes are applied automatically. It is possible for the driver to cancel the hooter but take no further action, and this has occurred on a few occasions. On busy lines drivers are constantly passing signals at yellow and cancelling the alarm can become almost a matter of course.

On London Transport the brakes are normally applied automatically if a train passes a signal at danger (red), but the Moorgate accident occurred when a driver approached a dead-end line at speed. More sophisticated control systems were suggested in the official report. [12] The psychological mechanisms that may have caused the driver to act as he did are dicussed by Reason and Mycielska. [13] From an engineering viewpoint, however, it is sufficient to realise that for one reason or another there is a significant chance (see Section 6.8) of a driver error and we must either accept the occasional error or prevent it by design.

2.7.2 Aircraft

Many aircraft have crashed because the pilots pulled a lever the wrong way. [14] For example, most modern jets are fitted with ground spoilers, flat metal plates hinged to the upper surface of each wing, which are raised *after* touch-down to reduce lift. They must not be raised *before* touch-down or the aircraft will drop suddenly.

On the DC-8 the pilot could either:

(a) *Lift* a lever before touch-down to arm the spoilers; they would then lift automatically after touch-down, or

(b) Wait until after touch-down and *pull* the same lever.

One day a pilot *pulled* the lever before touch-down. Result: 109 people killed.

The accident was not the fault of the pilot. It was the result of bad design. It was inevitable that sooner or later someone would move the lever the wrong way.

The reaction of the US Federal Aviation Administration was to suggest putting a notice in each cockpit alongside the spoiler lever saying "Deployment in Flight Prohibited". They might just as well have put up a notice saying "Do Not Crash this Plane".

The manufacturer of the DC-8, McDonnell Douglas, realised the notice was useless but wanted to do nothing. After two, perhaps three, more planes had crashed in the same way they agreed to fit locks to prevent the ground spoilers being raised before touch-down.

Many other aircraft accidents are blamed on pilot error when they could have been prevented by better design. Hurst writes:

"Some 60% of all accidents involve major factors which can be dismissed as 'pilot error'. This sort of diagnosis gives a . . . feeling of self-righteousness to those who work on the ground; but I want to state categorically that I do not believe in pilot error as a major cause of accidents. There are, it is true, a very few rare cases where it seems clear that the pilot wilfully ignored proper procedures and got himself into a situation which led to an accident. But this sort of thing perhaps accounts for one or two per cent of accidents – not 60%. Pilot error accidents occur, not because they have been sloppy, careless, or wilfully disobedient, but because we on the ground have laid booby traps for them, into which they have finally fallen."[15]

Many other slips, mostly trivial in their consequences are described by Reason and Mycielska. [1]

2.7.3 Roads

An "Accident Spotlight" issued jointly by a Police Force and a local Radio Station gave a diagram of a road junction where seven injury-causing accidents occurred in one year and then said,

"Principal cause : road user error".

There is no reason to believe that road users behave any worse at this junction than at any other and little hope that they will change their ways. Drawing attention to the number of accidents that occur at the junction may persuade a few people to approach it more cautiously but it would be better to redesign the junction. That however is expensive. It is cheaper to blame road users.

2.7.4 Mechanical handling

A plate fell from a clamp while being lifted because the screw holding it in position had not been tightened sufficiently (Figure 2.14 (a)). The operator knew what to do, but for some reason did not do it. The accident was put down as due to human failing and the operator was told to be more careful in future.

It would have been better to use a type of clamp such as that shown in Figure 2.14(b) which is not dependent for correct operation on someone tightening it up to the full extent. The accident could be prevented by better design.

2.8 Everyday life

I am indebted to my former secretary, Eileen Turner, for the following incident.

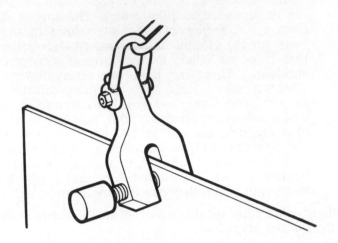

(a)   The screw must be tightened to the right extent

(b)   This design is not
        dependent on someone
        tightening a screw to
        the correct extent

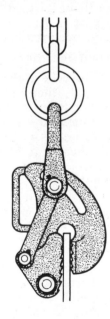

Figure 2.14   Two methods of attaching a metal plate to a hoist.

In an unusually houseproud mood, she cleaned the bedroom before going early to bed one night. She woke the next morning at six o'clock and, finding she couldn't get back to sleep, decided to get up and wash her hair.

After showering, brushing her teeth and washing her hair, she went into the living room, where after a few minutes, she noticed that the time by the rather old clock there was ten past one. The clock had obviously had its day and was going haywire but Eileen went to the bedroom to check. On first glance the time was now twenty to seven but closer examination showed that the clock was upside down! (Figure 2.15).

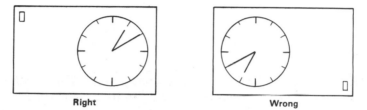

Right   **Wrong**

Figure 2.15   Two ways of putting a clock down.

To prevent a similar incident happening again she could change the hardware — modify the top of the clock so that it could not be put down upside down — or change the software — that is, give up dusting!

The following two 'human failing' accidents occurred during demonstration lectures.

> "The experimenter demonstrated the power of nitric acid to the subject by throwing a penny into it. The penny, of course, was completely disintegrated . . . While the subject's view of the bowl of acid was cut off by the experimenter, an assistant substituted for it a like-sized bowl of . . . water . . .
>
> The hypnotized subject was then ordered to throw the dish of nitric acid (in actual fact, of course, innocuous water) over the assistant who was present in the same room. Under these conditions it was possible to induce, under hypnosis, various subjects to throw what they considered to be an extremely dangerous acid into the face of a human being . . . Actually, in this particular experiment the person in charge made what he calls a 'most regrettable mistake in technique' by forgetting to change the nitric acid to the innocuous dish of water, so that in one case the assistant had real nitric acid thrown over him".[16]

29

"A deplorable accident has taken place at the Grenoble Lycée. The professor of chemistry was lecturing on salts of mercury, and had by his side a glass full of a mercurial solution. In a moment of distraction he emptied it, believing he was drinking a glass of eau sucrée. The unfortunate lecturer died almost immediately".[17]

The first accident could be prevented by not indulging in such a foolish experiment, which can so easily and obviously go wrong, and the second by using different containers for drinks and laboratory chemicals.

References to Chapter 2

1. J Reason and C Mycielska, *Absent-Minded? The Psychology of Mental Lapses and Everyday Errors,* Prentice-Hall, 1982.

2. Technical Data Note 46, *Safety at Quick-opening and other Doors of Autoclaves,* Factory Inspectorate, 1974.

3. Guidance Note FM/4, *High Temperature Dyeing Machines,* HMSO, 1976.

4. T A Kletz, *Loss Prevention,* Vol 13, 1980, p 1.

5. R T C Rolt, *Red for Danger,* Pan Books, 3rd edition, 1978.

6. A Schneider and A Masé, *Railway Accidents of Great Britain and Europe,* David and Charles, 1968.

7. J A B Hamilton, *Trains to Nowhere,* Allen and Unwin, 2nd edition, 1981.

8. M Gerard and J A B Hamilton, *Rails to Disaster,* Allen and Unwin, 1984.

9. J A B Hamilton, *Britain's Greatest Rail Disaster,* Allen and Unwin, 1969.

10. P M Olver, *Railway Accident: Report on the Collision that occurred on 14th February 1979 at Chinley North Junction in the London Midland Region of British Railways,* HMSO, 1981.

11. D R Davis, *Ergonomics,* Vol 9, 1966, p 211.

12. I K A McNaughton, *Railway Accident: Report on the Accident that occurred on 28th February 1975 at Moorgate Station on the Northern Line, London Transport Railways,* HMSO, 1976.

13. As Reference 1, p 204.

14. P Eddy, E Potter and B Page, *Destination Disaster,* Hart-Davis and MacGibbon, 1976.

15. R Hurst (Editor), *Pilot Error,* Crosby, Lockwood, Staples, 1976..

16. H J Eysenck, *Sense and Nonsense in Psychology,* Penguin Books, 1957.

17. *Nature,* 18 March 1880, quoted in *Nature,* 20 March 1980, p 216.

CHAPTER 3

ACCIDENTS THAT COULD BE PREVENTED BY BETTER TRAINING OR INSTRUCTIONS

I am under orders myself, with soldiers under me. I say to one, Go! and he goes; to another, Come here! and he comes; and to my servant, Do this! and he does it — A Centurion in St Matthew, chapter 8, verse 9 (New English Bible).

3.1 Introduction

This may have been true at one time, it may still be true in some parts of the world, but it is no longer true in Western industry and I doubt if it is still true in the Army. Tasks have changed and people's expectations have changed.

Tasks are no longer simple. We cannot lay down detailed instructions to cover every contingency. Instead we have to allow people judgement and discretion and we need to give them the skills and knowledge they need to exercise that judgement and discretion. We shall look at some accidents that occurred because people were not adequately trained.

People's expectations have changed. They are no longer content to do what they are told just because the boss tells them to. Instead they want to be convinced that it is the right thing to do. We need to explain our rules and procedures to those who will have to carry them out, and discuss them with them, so that we can understand and overcome their difficulties.

Of course, this argument must not be carried too far. Nine of your staff may agree on a course of action. The tenth man may never agree. He may have to be told, "We have heard your views. The rest of us have agreed. Now bloody well get on with it."

Sometimes accidents have occurred because of a lack of sophisticated training, as at Three Mile Island; sometimes because of a lack of basic training.

3.2 Three Mile Island

The accident at Three Mile Island Nuclear power station in 1979 had many causes and many lessons can be drawn from it [1,2] but some of the most important are concerned with the human factors. In particular, the training the operators had received had not equipped them to deal with the events that occurred.

32

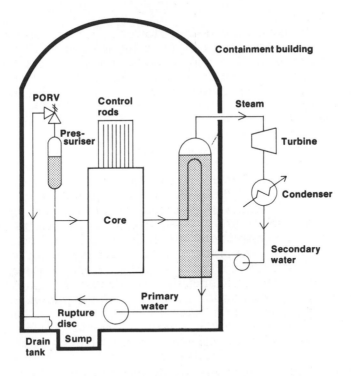

Figure 3.1 A pressurised water reactor – simplified.

To understand these I must briefly describe the design of the pressurised water reactor of the type used at Three Mile Island, and the events of 20 March 1979.

Figure 3.1 shows a very simplified reactor. Heat generated in the core by radioactive fission is removed by pumping primary water round and round it. This water is kept under pressure to prevent it boiling (hence it is called a pressurised water reactor, to distinguish it from a boiling water reactor). The primary water gives up its heat to the secondary water which does boil. The resulting steam then drives a turbine, before being condensed and the condensate recycled. All the radioactive materials, including the primary water, are enclosed in a containment building so that they will not escape if there is a leak.

The trouble started when a choke occurred in a resin polisher unit, which removes impurities from the secondary water. To try to clear the choke the operators used instrument air – at a lower pressure – so that water got back into the instrument air lines and the turbine tripped. This stopped the heat being removed from the radioactive core. The production of fission heat stopped automatically within a few minutes because silver rods which absorb neutrons dropped down into the core and stopped the radioactive fission. However, heat was still produced by radioactive decay at about 6% of the full load, and this caused the primary water to boil. The pilot operated relief valve (PORV) lifted and the make-up pumps started up automatically to replace the water which had evaporated from the primary circuit. *Unfortunately the PORV stuck open.* The operators did not realise this because a light on the panel told them it was shut. However, this light was not operated by the valve position but by the signal to the valve. The operators had not been told this or had forgotten.

Several other readings should have suggested to the operators that the PORV was stuck open and that the primary water was boiling. However, they chose to believe the PORV light and ignore the other readings, partly because they did not really understand how the temperature and pressure in the primary circuit depended on each other, and partly because their instructions and training had emphasised that it was dangerous to allow the primary circuit to get too full of water.

The operators thought the PORV was shut. Conditions were clearly wrong and their training had emphasised the danger of adding too much water. *They therefore shut down the make-up water pumps.*

Note that the only action taken by the operators made matters worse. If they had done *nothing* the system would have cooled down safely on its own. With the make-up water isolated, however, the level in the primary circuit fell and damage occurred.

The training of the operators was deficient in three major respects.

(1) It did not give them an understanding of the phenomena taking place in the primary water circuit; in particular, as already stated, they did not understand how the pressure and temperature were related;

(2) It did not tell them how to recognise a small loss of water – though it covered a major loss such as would occur if there was a break in the primary circuit – and what action to take if this occurred; and

(3) It did not train them in the skills of diagnosis. We cannot foresee everything that will go wrong and write instructions accordingly – though what did go wrong should have been foreseen – and so we need to train operators to diagnose previously unforeseen events.

One successful method of doing so has been described by Duncan and co-workers.[3] The operator is shown a mock-up of the control room panel on which various readings are displayed. From them he has to diagnose the fault and say what action he would take. The problems gradually increase in difficulty. The operator learns how to handle all foreseeable problems and acquires general skills which will help him handle unforeseen problems.

3.3 Other accidents that could be prevented by relatively sophisticated training.

The training required may not seem very sophisticated until we compare the incidents described in Section 3.4.

3.3.1 Re-starting a stirrer

A number of accidents have occurred because an operator found that a stirrer (or circulation pump) had stopped and switched it on again. Reactants mixed suddenly with violent results.

For example, an acidic effluent was neutralised with a chalk slurry in a tank. The operator realised that the effluent going to drain was too acidic. On looking round, he found that the stirrer had stopped. He switched it on again. The acid and the chalk reacted violently, blew off the manhole cover on the tank and lifted the bolted lid. No-one was injured.

A similar incident had occurred on the same plant about four years earlier. An instruction was then issued detailing the action to be taken if the stirrer stopped. The operator had not seen the instruction or, if he had seen it, he had forgotten it. No copy of it could be found on the plant.

It is difficult to prevent accidents such as this by a change in design. An alarm to indicate that the stirrer had stopped might help and the manhole on the tank should be replaced by a hinged lid which will lift when the pressure rises but it will still be necessary to train the operators and maintain the instructions. Initial training is insufficient. Regular refreshers are necessary. It is also useful to supplement training with a plant 'black book', a folder of reports on incidents that have occurred. It should be compulsory reading for newcomers and it should be kept in the control room so that others can dip into it in odd moments. It should not be cluttered up with reports on tripping accidents and other trivia, but should contain reports on incidents of technical interest, both from the plant and other similar plants.

Instructions, like labels, are a sort of protective system and like all protective systems they should be checked regularly to see that they are complete and maintained as necessary.

3.3.2 Clearing choked lines

Several incidents have occurred because people did not appreciate the power of gases or liquids under pressure and used them to clear chokes.

For example, high pressure water wash equipment was being used to clear a choked line. Part of the line was cleared successfully, but one section remained choked so the operators decided to connect the high pressure water directly to the pipe. As the pressure of the water was 100 bar (it can get as high as 650 bar) and as the pipe was designed for only about 10 bar, it is not surprising that two joints blew. Instead of suspecting that something might be wrong the operators had the joints remade and tried again. This time a valve broke.

Everyone should know the safe working pressure of their equipment and should never connect up a source of higher pressure without proper authorisation by a professional engineer who should first check that the relief system is adequate.

On another occasion gas at a gauge pressure of 3 bar — which does not seem very high — was used to clear a choke in a 2 inch pipeline. The plug of solid was moved along with such force that when it hit a slip-plate it made it concave. Calculation, neglecting friction, showed that if the plug weighed 0.5 kg and it moved 15 m, then its exit velocity would be 500 km/hour!

An instrument mechanic was trying to free, with compressed air, a sphere which was stuck inside the pig chamber of a meter prover. Instead of securing the chamber door properly he fixed it by inserting a metal rod — a wheel dog — into the top lugs. When the gauge pressure reached 7 bar the door flew off and the sphere travelled 230 m before coming to rest, hitting various objects on the way.[4]

In all these cases it is clear that the people concerned had no idea of the power of liquids and gases under pressure. Many operators find it hard to believe that a 'puff of air' can damage steel equipment.

In the incident described in Section 2.2.1, Example 1, the other operators on the plant found it hard to believe that a pressure of "only 30 pounds" caused the door of the filter to fly open with such force. They wondered if a chemical explosion had occurred.

The operators did not understand the difference between force and pressure. They did not understand that a force of 30 pounds was exerted against every square inch of a door that was 3.5 feet in diameter and that the total force on the door was therefore 20 tons.

3.3.3   Failures to apply well-known knowledge

It is not, of course, sufficient to have knowledge. It is necessary to be able to apply it to real-life problems. Many people seem to find it difficult. They omit to apply the most elementary knowledge. For example, scaffolding was erected around a 75 m tall distillation column so that it could be painted. The scaffolding was erected when the column was hot and then everyone was surprised that the scaffolding became distorted when the column cooled down.[5]

Many pressure vessels have burst when exposed to fire. For example, at Feyzin in France in 1966 a 2000 m^3 sphere containing propane burst, killing 18 people and injuring many more.[6-10] The vessel had been exposed to fire for 1½ hours before it burst and during this time the

36

Fire Brigade had, on the advice of the refinery staff, used the available water for cooling surrounding vessels to prevent the fire spreading. The relief valve, it was believed, would prevent the vessel bursting.

I have discussed the Feyzin fire on many occasions with groups of students and with groups of experienced operating staff and their reaction is often the same: The relief valve must have been faulty or too small or there must have been a blockage in its inlet or exit pipe. When they are assured that the relief valve was OK it may still take them some time to realise that the vessel burst because the metal got too hot and lost its strength. Below the liquid level the boiling liquid kept the metal cool, but above the liquid level the metal softened and burst.

Everyone knows that metal loses its strength when it becomes hot, but there was a failure to apply that knowledge, both by the refinery staff at the time and by the people who attended my discussions.

It is fair to state that today people realise why the sphere burst much more quickly than they did in the past.

Failure to apply the knowledge we have is not, of course, a problem peculiar to plant operators. One of the major problems in education is not giving knowledge to people, but persuading them to use it. Most of us keep "book learning" and "real life" in separate compartments and the two rarely meet. One method of helping to break down the barrier between them is by the discussion of incidents such as Feyzin. The group puzzle out why it occurred and say what *they think* should be done. The Institution of Chemical Engineers provide sets of notes and slides for use in this way.[11]

3.3.4 Contractors

Many accidents have occurred because contractors were not adequately trained.

For example, storage tanks are usually made with a weak seam roof, so that if the tank is overpressured the wall/roof seam will fail rather than the wall/floor seam. On occasions contractors have strengthened the wall/roof seam, not realising that it was supposed to be left weak.

Many pipe failures have occurred because contractors failed to follow the design in detail or to do well what was left to their discretion.[12] The remedy lies in better inspection after construction, but is it possible to give contractors' employees more training in the *consequences* of poor workmanship or short cuts on their part? Many of them do not realise the nature of the materials that will go through the completed pipelines and the fact that leaks may result in fires, explosion, poisoning or chemical burns. Many engineers are sceptical of the value of such training. The typical construction worker, they say, is not interested in such things. Nevertheless, it might perhaps be tried. (Construction errors are discussed further in Chapter 8.)

3.3.5   Information on change

Accidents have occurred because changes in design necessitated changes in construction or operating methods, but those concerned were not told. Our first example concerns construction.

In Melbourne, Australia in 1970, a box girder bridge, under construction across the Yarra river, collapsed during construction. The cause was not errors in design, but errors in construction: an attempt to force together components that had been badly made and did not fit.

However it is customary for construction teams to force together components that do not fit. No-one told them that with a box girder bridge – then a new type – components must not be forced together; if they do not fit they must be modified. The consulting engineers made no attempt to ensure that the contractors understood the design philosophy and that traditional methods of construction could not be used. Nor did they check the construction to see that it was carried out with sufficient care.[13]

A 25 ton telescopic jib crane was being used to remove a relief valve, weight 258 lb, from a plant. The jib length was 124 feet and the maximum safe radius for this jib length is 80 feet. The driver increased the radius to 102 feet and the crane fell over onto the plant (Figure 3.2).

Figure 3.2   This crane tried to lift too great a load for the jib length and elevation.

Damage was slight but the plant had to be shut down and depressured while the crane was removed. The crane was fitted with a safe load indicator of the type that weighs the load through the pulley on the hoist rope; it does not take into account the weight of the jib. Because of this the driver got no warning of an unsafe condition and, as he lifted the valve, the crane overturned. The driver had been driving telescopic jib cranes for several years but did not appreciate the need not to exceed the maximum jib radius. He did not realise that the crane could be manoeuvred into an unstable position without an alarm sounding.

Those responsible for training the driver had perhaps failed to realise themselves that the extra degree of freedom on a telescopic jib crane — the ability to lengthen the jib — means that it is easier to manoeuvre the crane into an unsafe position. Certainly they had failed to take this into account in the training of the driver. They had, incidentally, also failed to realise that an extra degree of freedom requires a change in the method of measuring the approach to an unsafe condition.

In the summer of 1974, a plastic gas main was laid in a street in Freemont, Nebraska alongside a hotel, and was fixed to metal mains at each end by compression couplings. In the following winter the pipe contracted and nearly pulled itself out of the couplings. The next winter was colder and the pipe came right out. Gas leaked into the basement of the hotel and exploded. The pipe was 348 feet long and contracted about 3 inches.[14]

Apparently nobody told the men who installed the pipe that when plastic pipe is used in place of metal pipe, it is necessary to allow for contraction. (This might be classified as an accident due to failure to apply well-known knowledge. Everyone knows that substances contract on cooling.)

At a more prosaic level, accidents have occurred because people were not told of changes made while they were away. An effluent had to be neutralised before it left a plant. Sometimes acid had to be added, sometimes alkali. Sulphuric acid and caustic soda solution were supplied in similar plastic containers (polycrates). The acid was kept on one side of the plant and the alkali on the other. While an operator was on his days off someone decided it would be more convenient to have a container of acid and a container of alkali on each side of the plant. When the operator came back no-one told him about the change. Without checking the labels he poured some excess acid into a caustic soda container. There was a violent reaction and he was sprayed in the face. Fortunately he was wearing goggles [15] (Figure 3.3).

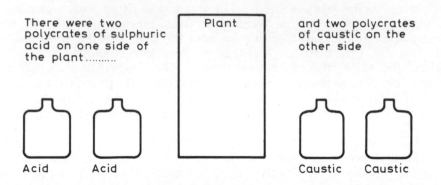

There were two polycrates of sulphuric acid on one side of the plant..........

Plant

and two polycrates of caustic on the other side

Acid   Acid   Caustic   Caustic

While an operator was on his days off someone decided it would be more convenient to have a polycrate of acid and a polycrate of alkali on each side.

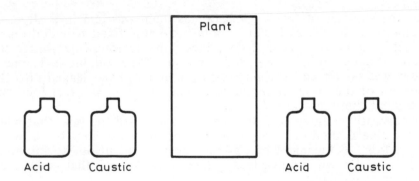

Plant

Acid   Caustic   Acid   Caustic

Figure 3.3   Arrangement of acid and caustic containers on a plant.

3.3.6   Team working

With the reduction in manpower on many plants it has become fashionable to employ 'team working'. The operators are not assigned specific tasks but each man is capable of carrying out the whole range of operations and each task is done by whoever is most conveniently available at the time.

This obviously encourages efficiency, gives people a larger range of interests and sounds good, but there are snags:

— Extensive training is necessary to make sure that every man can really carry out the full range of tasks.

—   A situation in which anyone can do a job, can easily become a situation in which no-one does it. Everybody thinks someone else is doing it. A system is needed for making sure that this does not occur.

One incident occurred (in part) because a control room operator, over the loudspeaker, asked the outside operators to close a valve. There were two outside operators who worked as a team. The control room operator did not ask a particular outside man to close the valve, both were some distance away, and each left it to the other whom he thought would be nearer.

3.4   Accidents that could be prevented by elementary training

This section describes some incidents which occurred because people lacked the most elementary knowledge of their duties or the properties of the material they handled.

(a)   Many accidents have occurred because operators wrote down abnormal readings on the plant record sheets but did nothing about them. They did not even tell their foremen. They thought their job was just to take readings, not to respond to them. An example is described in Section 4.4.2.

(b)   After changing a chlorine cylinder, two men opened valves to make sure there were no leaks on the lines leading from the cylinder. They did not expect to find any so they did not wear breathing apparatus. Unfortunately there were some and they were affected by the chlorine.

   If the men were sure there would be no leaks, there was no need to test. If there was a need to test, then leaks were possible and breathing apparatus should have been worn.

(c)   A liquid was transferred into a tank by vacuum. The tank was emptied with a vacuum pump and then the pump was shut down and isolated and the liquid drawn into the tank. This was not quick enough for one shift who kept the pump running. Someone told them that spray from the inflow would be drawn into the pump. The operators said "No, it doesn't, you see, all the time the pump is running there's a strong stream of 'vac' running into the the tank and it keeps the splashes away". [16] It is fair to point out that this incident occurred some years ago.

(d)   An operator had to empty tank wagons by gravity. His instructions said:

   1. Open the vent valve on the top of the wagon.

   2. Open the drain valve.

   3. When the tanker is empty, close the vent valve.

One day he had an idea. To save climbing onto the top of the wagon twice, he decided to carry out step 3 before step 2.

Result: the wagon was sucked in.

The operator did not understand that when a liquid flows out of a tank it leaves an empty space behind.

(e)   Petrol pump attendants seem to be the most untrained of all employees, despite the hazardous nature of the material they handle. For example:

− A man got out of a mini-van just in time to prevent the attendant filling it up through the ventilator in the roof.

− A young attendant used his cigarette lighter to check the level in a road tanker.

− An attendant put petrol into the oil filler pipe of a rear-engined van. The engine caught fire when the ignition was switched on.

In this last case the attendant was watched by the driver who himself removed the oil filler cap.[17]

(f)   Maintenance workers as well as operators occasionally display the most extraordinary ignorance of what practices are acceptable.

A 24 inch manhole branch on a vessel had corroded and had to be replaced. A new branch was made and the old one removed. When the new branch was offered up to the opening in the vessel it was found to be a little bit too small; the old branch was 24 inches *internal* diameter, while the new one was 24 inches *external* diameter. The supervisor therefore decided to make a series of parallel cuts in the branch, splay them out until they fitted the hole in the vessel and weld up the gaps! Fortunately before the job was complete it came to the notice of a senior engineer and was stopped.

Other incidents are described in Chapter 9.

In a sense all the incidents described in this Chapter were due to human failing but it is not very helpful to say so. The incidents could have been prevented by better management, in these cases by better training and instructions.

3.5   Some simple ways of improving instructions

While training is concerned with giving people the knowledge and understanding they need to do their job properly, instructions are concerned with specific tasks and the precise course of action to be followed.

As well as the well-established methods of training, there are newer methods touched on in Sections 3.2 and 3.3.3. Below we list some questions that should be asked about instructions.

— Are they easy to read?

— Are they explained to those who will have to carry them out?

— Are they maintained?

3.5.1   Are they easy to read?

Many instructions are not. Figure 3.4 is typical of the language and appearance of many. Figure 3.5 shows a much better layout.

Men are remarkably good at detecting meaning in a smog of verbiage but they should not be expected to do so. Sooner or later they will fail to comprehend.

One of the advantages of computer control is that it forces us to give precise instructions.

3.5.2   Are they explained to those who will have to carry them out?

On one Works the instruction on the procedure to be followed and the precautions to be taken before men were allowed to enter a vessel or other confined space ran to 23 pages plus 33 pages of appendices, 56 pages in all. There were many special circumstances but even so this seems rather too long. However, when the instruction was revised it was discussed in draft with groups of supervisors and the changes pointed out. This was time-consuming for both supervisors and managers but was the only way of making sure that the supervisors understood the changes and for the managers to find out if the changes would work in practice.

Most people, on receiving a 56 page document, will put it aside to read when they have time — and you know what that means. New instructions should be discussed with those who will have to carry them out.

3.5.3   Are they maintained?

Necessary maintenance is of two sorts. First, the instructions must be kept up-to-date and, second, regular checks should be made to see that the instructions are present in the control room and in a legible condition. If too worn they should obviously be replaced. If spotlessly clean, like poetry books in libraries, it suggests that they are never read and the reasons for this should be sought. Perhaps they are incomprehensible.

A senior manager visiting a control room should ask to see the instructions — operating as well as safety. He may be surprised how

```
INSTRUCTION NO:      WC 101

TITLE:               HOW TO LAY OUT OPERATING INSTRUCTIONS
                     SO THAT THEY MAY BE READILY DIGESTED
                     BY PLANT OPERATING STAFF.

AUTHOR:              EAST SECTION MANAGER

DATE:                1 DECEMBER 1976

COPIES TO:           UNCLE TOM COBBLEY AND ALL


Firstly, consider whether you have considered every eventuality
so that if at any time in the future anyone should make a mistake
whilst operating one of the plants on East Section you will be able
to point to a piece of paper that few people will know exists and
no-one other than yourself will have read or understood.  Don't use
one word when five will do, be meticulous in your use of the
English language and at all times ensure that you make every
endeavour to add to the vocabulary of your operating staff by using
words with which they are unfamiliar, for example, never start
anything, always initiate it.  Remember that the man reading this
has turned to the instructions in desperation, all else having
failed, and therefore this is a good time to introduce the
maximum amount of new knowledge.  Don't use words, use numbers,
being careful to avoid explanations or visual displays, which would
make their meaning rapidly clear.  Make him work at it; it's a
good way to learn.

Wherever possible use the instruction folder as an initiative test;
put the last numbered instruction first, do not use any logic in the
indexing system, include as much information as possible on
administration, maintenance data, routine tests, plants which are
geographically close and training randomly distributed throughout
the folder so that useful data is well hidden, particularly that
which you need when the lights have gone out following a power
failure.
```

Figure 3.4   **ARE YOUR PLANT INSTRUCTIONS LIKE THIS?**

The following extract from a plant instruction shows the action a supervisor and four operators should take when the induced draught fan providing air to a row of furnaces (known as A side) stops. Compare the lay-out with that of Figure 3.4.

ACTION TO TAKE WHEN A SIDE ID FAN TRIPS

1   CHECK A SIDE FURNACES HAVE TRIPPED

2   ADJUST KICK-BACK ON COMPRESSORS TO PREVENT SURGING

3   REDUCE CONVERTER TEMPERATURES

4   CHECK LEVEL IN STEAM DRUMS TO PREVENT CARRY-OVER

Panel Operator

1   Shut TRC's on manual

2   Reduce feed rate to affected furnaces

3   Increase feed to Z furnace

4   Check temperature of E54 column

Furnace Operator

1   Fire up B side and Z furnaces

2   Isolate liquid fuel to A side furnaces

3   Change over superheater to B side

4   Check that output from Z furnace goes to B side

Centre Section Operator

1   Change pumps onto electric drive

2   Shut down J43 pumps

Distillation Operator

1   Isolate extraction steam on compressor

2   Change pumps onto electric drive

Figure 3.5 — **OR LIKE THIS**

often they are out-of-date, or cannot readily be found or are spotlessly clean.

I know of one explosion which occurred because an operator followed out-of-date instructions he found in a folder in the control room.

Finally, a quotation from H J Sandvig:

> "Operators are taught by other operators and . . . . each time this happens something is left unsaid or untold unless specific operating instructions are provided, specific tasks are identified and written and management reviews these procedures at least annually and incorporates changes and improvements in the process". [18]

Some incidents which could have been prevented by better instructions are described in Section 10.3.

3.6 Cases when training may not be the best answer

3.6.1 Electric plugs

Ordinary domestic 3-pin electric plugs are sometimes wired incorrectly. This may be due to a slip, or perhaps to ignorance of the correct method. The usual remedy is training and instruction: electric appliances come with instructions for fitting the plug. However, it is not difficult to reduce the opportunities for error. One method, which is on the increase, is to fit the plug in the factory. Another method, so simple it is surprising that it has not been used, is to colour the terminals in the plug the same colour as the wires.

3.6.2 Kinetic handling

Training in kinetic handling methods is often recommended as a cure for back injuries but in many cases it is only part of the story. Before introducing a training programme the layout and design of areas where people have to work should be examined to check that they are the best possible. For example, a man has been seen working with his feet at different levels, lifting 30 lb from a conveyor belt onto staging behind him. A hoist was needed, not training of the man.

In another example there were two almost identical conveyor layouts, yet back accidents occurred on one and never on the other. An examination showed the first line was close to very large doors so that, when men stopped for a rest after getting hot and sweaty, they were standing in a draught, with the not unnatural result that they suffered back trouble.

In another example, trucks brought loads from stores and deposited them in unoccupied areas near to but not at the actual place where the goods would be required, with the result that men were then called upon to manhandle them over the last part. Very often the temporary resting places of these goods were pebbled areas and other unsatisfactory places so that men did not have a proper footing, with inevitable falls and strains as a result.

In short, we should examine the layout and planning of a task before considering training.

3.6.3   Attitudes

It is sometimes suggested that through training we should try to change people's attitudes.

I doubt if such training is either justified or effective.

It is not justified because a person's attitude is his private affair. We should concern ourself only with whether or not he achieves his objectives.

It is not effective because it is based on the assumption that if we change a person's attitude we change his behaviour. In fact, it is the other way round. An attitude has no real existence; it is just a generalisation about behaviour.

If someone has too many accidents, let us discuss the reasons for them and the action needed to prevent them happening again. After a while he may act differently, he may have fewer accidents and everyone will say that he has changed his attitude.
In short:

Don't try to change people's attitudes

Just help them with their problems

William Blake wrote, "He who would do good to another must do it in Minute Particulars. General Good is the plea of the scoundrel, hypocrite and flatterer".[19] See Section 5.2.5.

Similarly, do not try to persuade Boards of Directors to change their policies. It is better to suggest ways of dealing with specific problems. Looking back, a common pattern may be seen. This is the policy — the common law of the organisation.

References to Chapter 3

1.   *Report of the President's Commission on the Accident at Three Mile Island (The Kemeny Report),* Pergamon Press, 1979.

2.   T A Kletz, *Hydrocarbon Processing,* Vol 61, No 6, June 1982, p 187.

3. E E Marshall et al, *The Chemical Engineer,* No 365, Feb 1981, p 66.

4. *Petroleum Review,* July 1983, p 27.

5. *Petroleum Review,* July 1982, p 21.

6. *The Engineer,* 25 March 1966, p 475.

7. *Paris Match,* No 875, 15 Jan 1966.

8. *Fire,* Special Supplement, Feb 1966.

9. *Petroleum Times,* 21 Jan 1966, p 132.

10. P Lagadec, *Major Technological Risks,* Pergamon Press, 1980, p 175.

11. *Hazard Workshop Modules,* Institution of Chemical Engineers, 5 sets.

12. T A Kletz, *Plant/Operations Progress,* Vol 3, No 1, Jan 1984, p 19.

13. *Report of Royal Commission into the Failure of the West Gate Bridge,* State of Victoria Government Printer, Melbourne, 1971.

14. From a report issued by the US National Transportation Safety Board, 1975.

15. *Petroleum Review,* Jan 1974.

16. A Howarth, *Chemistry in Britain,* Vol 20, No 2, Feb 1984, p 140.

17. *The Guardian,* 23 June 1971.

18. H J Sandvig, *JAOCS,* Vol 60, No 2, Feb 1983, p 243.

19. W Blake, *Jerusalem,* 55, 60.

CHAPTER 4

ACCIDENTS DUE TO A LACK OF PHYSICAL
OR MENTAL ABILITY

*Nothing is impossible for people who do not
have to do it themselves* − Anon.

These accidents are much less common than those described in Chapters 2 and 3 but nevertheless do occur from time to time. By the title I do not intend to imply that a significant number of accidents occur because of individual lack of ability, but rather that some occur because people are asked to do more than people as a whole are capable of doing, physically or mentally. Only a few occur because the individual concerned was asked to do more than he (or she) was individually capable of doing.

4.1 People asked to do the physically difficult or impossible

(a) A steel company found that overhead travelling magnet cranes were frequently damaging railway wagons. One of the causes was found to be the design of the crane cab. The driver had to lean over the side to see his load. He could then not reach one of the controllers, so he could not operate this control and watch the load at the same time.[1]

(b) A refinery compressor was isolated for repair and swept out with nitrogen but as some hydrogen sulphide might still be present, the fitters were told to wear air-line breathing apparatus. They found it difficult to remove a cylinder valve which was situated close to the floor, so one fitter decided to remove his mask and was overcome by the hydrogen sulphide. Following the incident lifting aids were provided.

   Many companies would have been content to reprimand the fitter for breaking the rules. The company concerned, however, asked why he had removed his mask and it then became clear that he had been asked to carry out a task which was difficult to perform while wearing a mask.[2]

(c) Incidents have occurred because valves which have to be operated in an emergency were found to be too stiff. Such valves should be kept lubricated and exercised from time to time.

(d) Operators often complain that valves are inaccessible. Emergency valves should, of course, always be readily accessible but other valves, if they have to be operated, say, once/year or less often, can be out of reach. It is reasonable to expect operators to fetch a ladder or scramble into a pipe trench at this frequency.

Designers should remember that if a valve is just within reach of an average person then half the population cannot reach it. They should design so that 95% (say) of the population can reach it.

(e)   It is difficult to give lorry drivers a good view while they are reversing. Aids such as large mirrors should be provided in places such as loading bays where a lot of reversing has to be done.[1]

(f)   Related to these accidents are those caused by so-called clumsiness. For example, an electrician was using a clip-on ammeter inside a live motor starter cubicle. A clip-on ammeter has two spring-loaded jaws which are clipped round a conductor forming a coil which measures the current by induction. The jaws are insulated except for the extreme ends.

The electrician accidentally shorted two live phases (or a live phase and earth) with the bare metal ends of the jaws. He was burnt on the face and neck and the starter was damaged.[3] In many companies such an accident would be put down to 'clumsiness' and the man told to take more care.

Such accidents are the physical equivalent of the mental slips discussed in Chapter 2. The method of working makes an occasional accident almost inevitable. Sooner or later, for one reason or another, the electrician's co-ordination of his movements will be a little poorer than normal and an accident will result. We all have off days. The method of working is hazardous — though accepted practice — and a better method should be sought.

4.2   People asked to do the mentally difficult or impossible

Many of the incidents described in Chapter 2 almost come into this category. If a man is told to close a valve when an alarm sounds or at a particular stage in an operation he cannot be expected to close the right valve in the required time *every time*. His error rate will be higher if the layout or labelling are poor or he is under stress or distracted. However he will close the right valve most of the time. In this section we consider some accidents which occurred because people were expected to carry out tasks which most people would fail to carry out correctly on many occasions. These failures are of several sorts: Those due to information or task overload or underload, those involving detection of rare events, those involving habits and those involving judgments.

4.2.1   Information or task overload

A new, highly-automated plant developed an unforeseen fault. The computer started to print out a long list of alarms. The operator did not know what had occurred and took no action. Ultimately an explosion occurred.

Afterwards the designers agreed that the situation should not have occurred and that it was difficult or impossible for the operator to diagnose the fault, but they then said to him, "Why didn't you assume the worst and trip the plant? Why didn't you say to yourself, 'I don't know what's happening so assume it is a condition that justifies an emergency shut-down. It can't be worse than that'?"

Unfortunately men do not work like that. If a man is overloaded with too much information he may simply switch-off and do nothing. The action suggested by the designers may be logical, but this is not how men behave under pressure.

The introduction of computers has made it much easier than in the past to overload people with too much information, in management as well as operating jobs. If quantities of computer print-out are dumped on people's desks every week, then most of it will be ignored, including the bits that should be looked at.

Plant supervisors sometimes suffer from task overload, that is, they are expected to handle more jobs at once than a person can reasonably cope with. This has caused several accidents. For example, two jobs had to be carried out simultaneously in the same pipe trench, 20 m apart. The first job was construction of a new pipeline. A permit-to-work, including a welding permit, was issued at 08.00 hours, valid for the whole day. At 12.00 hours a permit was requested for removal of a slip-plate from an oil line. The foreman gave permission, judging that the welders would by this time be more than 50 feet (17 m) from the site of the slip-plate. He did not visit the pipe trench which was 500 m away, as he was busy on the operating plant. Had he done so he might have noticed that the pipe trench was flooded. Although the pipeline had been emptied, a few gallons of light oil remained and ran out when the slip-plate joint was broken. It spread over the surface of the water in the pipe trench and was ignited by the welders. The man removing the slip-plate was killed.

The actual distance between the two jobs − 20 m − was rather close to the minimum distance − 17 m − normally required. However the 17 m includes a large safety margin. Vapour from a small spillage will not normally spread anything like this distance. On the surface of water, however, liquid will spread hundreds of metres.

Afterwards a special day supervisor was appointed to supervise the construction operations. It was realised that it was unrealistic to expect the foreman, with his primary commitment to the operating plant, to give the construction work the attention it required.

If extensive maintenance or construction work is to be carried out on an operating plant, for example, if part of it is shut down and the rest running, extra supervision should be provided. One supervisor should look after normal plant operations and the other should deal with the maintenance or construction organisation.

Overloading of a supervisor at a busy time of day may have contributed to another serious accident. At supervisor returned to work after his days off at 08.00 hours on Monday morning. One of his first jobs was to issue a permit-to-work for repair of a large pump. When the pump was dismantled, hot oil came out and caught fire. Three men were killed and the plant was destroyed. It was found that the suction valve on the pump had been left open.

The supervisor said he had checked the pump before issuing the permit and found the suction valve (and delivery valve) already shut. It is possible that it was and that someone later opened it. It is also possible that the supervisor, due to pressure of work, forgot to check the valve.

The real fault here, of course, is not the overloading of the supervisor — inevitable from time to time — but the lack of a proper system of work. The valves on the pump should have been locked shut and in addition the first job, when maintenance started, should have been to insert slip-plates. If valves have to be locked shut, then supervisors have to visit the scene in order to lock them.

Overloading of shift supervisors can be reduced by simple arrangements. They should not, for example, start work at the same time as the maintenance team.

4.2.2 Detection of rare events

If a man is asked to detect a rare event he may fail to notice when it occurs, or may not believe that it is genuine. The danger is greatest when he has little else to do. It is very difficult for night watchmen, for example, to remain alert when nothing has been known to happen and when there is nothing to occupy their minds and keep them alert. (Compare *St Matthew,* 26, 40, "Could ye not watch with me one hour?") On some of the occasions when train drivers have passed a signal at danger (see Section 2.7.1.2), the drivers, in many years of driving regularly along the same route, had never before known that signal to be at danger.

Similarly, in the days before continuously braked trains became the rule, railway signalmen were expected to confirm that each train that passed the signal box was complete, as shown by the presence of tail lights.

In 1936 an accident occurred because a signalman failed to spot the fact that there were no tail lights on the train. In his 25 years as a signalman he had never previously had an incomplete train pass his box.[4] The signalman was blamed (in part) for the accident but it is difficult to believe that this was justified. It is difficult for anyone to realise that an event that has never occurred in 25 years has in fact just happened.

In September 1984 the press reported that after a large number of false alarms had been received, an ambulance crew failed to respond to a call that two boys were trapped in a tank. The call was genuine. (See also Section 5.2.2.)

4.2.3   Task underload

It is widely recognised that reliability falls off when people have too little to do as well as when they have too much to do. As already stated, it is difficult for night watchmen to remain alert.

During the war studies were made of the performance of watch-keepers detecting submarines approaching ships. It was found that the effectiveness of a man carrying out such a passive task fell off very rapidly after about 30 minutes.

It is sometimes suggested that we should restrict the amount of automation on a plant in order to give the operators enough to do to keep them alert. I do not think this is the right philosophy. If automation is needed to give the necessary reliability, then we should not sacrifice reliability in order to find work for the operators. We should look for other ways of keeping them alert. Similarly if automation is chosen because it is more efficient or effective, we should not sacrifice efficiency or effectiveness in order to find work for the operators.

In practice, I doubt if process plant operators often suffer from task underload to an extent that affects their performance. Although in theory they have little to do on a highly automated plant, in practice there are often some instruments on manual control, there are non-automated tasks to be done, such as changing over pumps and tanks, there is equipment to be prepared for maintenance, routine inspections to be carried out, and so on.

If, however, it is felt that the operators are seriously under-loaded, then we should not ask them to do what a machine can do better but look for useful but not essential tasks that will keep them alert and which can be set aside if there is trouble on the plant — the process equivalent of leaving the ironing for the babysitter.

One such task is the calculation and graphing of process parameters such as efficiency, fuel consumption, catalyst life and so on.

Another task is studying a training programme such as a slide/tape sequence. This is best done under the supervision of a foreman who stops the programme at suitable intervals, adds extra explanation if necessary and discusses it with the operators.

Despite what has been said above, fully operated plants are not necessarily the most reliable. Hunns[5] has compared three designs for a boiler control system: A largely manual design, a partly-automated design and a fully automated design. The comparison showed that the partly-automatic design was the most reliable. Section 6.5.2 shows that an operator may be more reliable at some process tasks than automatic equipment.

More important than occupying the operator's time, is letting him feel that he is in charge of the plant — able to monitor it and able to intervene when he considers it necessary — and not just a passive bystander watching a fully automatic system. The control system should be designed with this philosophy in mind.[6]

4.2.4 Habits

If we expect people to go against established habits, errors will result. For example, if a man is trained to drive a vehicle with the controls laid out in one way and is then transferred to another vehicle where the layout is different, he will probably make mistakes. A good example occurred in a company which was developing an attachment to fit on the back of a tractor. The driving controls had to be placed behind the seat so that the driver could face the rear. On the development model the control rods were merely extended to the rear and thus the driver found that they were situated in the opposite positions to those on a normal tractor. When someone demonstrated the modified tractor, he twice drove it into another vehicle.[1]

Habits are also formed by general experience as well as experience on particular equipment. In the UK people expect that pressing a switch down will switch on a light or appliance. If a volume control is turned clockwise we expect the sound to get louder. If the controller on a crane is moved to the right we expect the load to move to the right. Errors will occur if designers expect people to break these habits.

4.2.5 Judgments of dimensions

People are not very good at estimating distances. Vehicles are often driven through openings which are too narrow or too short. Drivers should know the dimensions of their vehicles and the dimensions of openings should be clearly displayed. [1]

4.3   Individual traits and accident proneness

A few accidents occur because individuals are asked to do more than they are capable of doing, though the task would be within the capacity of most people.

For example, at a weekly safety meeting a maintenance foreman reported that one of his men had fallen off a bicycle. He had been sent to the store for an item that was required urgently and on the way, while entering a major road, had collided with another cyclist. The bicycle and road surface were in good condition, and the accident seemed to be a typical 'human failing' one.

Further inquiries, however, disclosed that the man was elderly, with poor eyesight, and unused to riding a bicycle. He had last ridden one when he was a boy. The foreman had sent him because no-one else was available.

In 1935 a signalman on a busy line accepted a second train before the first one had left the section. It came out in the inquiry that although he had 23 years' experience and had worked in four different boxes, he had been chosen for this present box by seniority rather than merit and had taken 5 weeks to learn how to work it.[4]

This brings us to the question of accident proneness which has been touched on already in Section 2.1. It is tempting to think that we can reduce accidents by psychological testing but such tests are not easy and few accidents seem to occur because individuals are accident-prone. Hunter[7] writes:

> "Accident proneness is greatly influenced by the mental attitude of the subjects. The accident-prone are apt to be insubordinate, temperamentally excitable and to show a tendency to get flustered in an emergency. These and other defects of personality indicate a lack of aptitude on the part of the subjects for their occupation. But whenever we are considering how to prevent accidents we must avoid the danger of laying too much blame on abnormalities of temperament and personality. Let us beware lest the concept of the accident-prone person be stretched beyond the limits within which it can be a fruitful idea. We should indeed be guilty of a grave error if for any reason we discouraged the manufacture of safe machinery."

Swain[8] describes several attempts to measure the contribution of accident-prone individuals to the accident rate. In one study of 104 railway shunters over three years, the ten shunters with the highest accident rate in the first year were removed from the data for the following two years. The accident rate for these two years actually rose slightly. Similar results were obtained in a study of 847 car drivers.

Eysenck[9], however, shows that personality testing of South African bus-drivers reduced accident rates by 60% over 10 years.

Whatever may be the case in some industries, in the process industries, very few individuals seem to have more than their fair share of accidents. If anyone does, we should obviously look at his work situation and if it is normal, consider whether he is suitable for the particular job. However, remember that a man may have more than the average share of accidents by chance. Suppose that in a factory of 675 men there are in a given period 370 accidents. Obviously there are not enough accidents to go round and many men will have no accidents and most of the others will have only one. If the accidents occur at random, then

11   men will have 3 accidents each

1.5   men will have 4 accidents each

0.15 men will have 5 accidents each,

that is, once in every 6 or 7 periods we should expect one man to have 5 accidents by chance.

Swain[8] describes a study of 2300 US railway employees. They were divided into 1828 low-accident men, who had four or less accidents, and 472 high-accident men, who had five or more accidents. If the accidents are distributed at random then we would expect 476 men to have five or more accidents.

Another reason why some men may have an unusually large number of accidents is that they deliberately but unconsciously injure themselves in order to have an excuse for withdrawing from a work situation which they find intolerable[10] (or, having accidentally injured themselves, use this as an excuse to withdraw from work).

Obviously changing the job so as to remove opportunities for accidents will not prevent accidents of this type. The man will easily find another way of injuring himself. If we believe that a man's injuries occur in this way then we have to try to find out why he finds work intolerable. Perhaps he does not get on with his fellow workers; perhaps his job does not provide opportunities for growth, achievement, responsibility and recognition. (See Appendix.)

To sum up on accident proneness, if a man has an unusually large number of accidents, compared with his fellows, may be due to:

(1)   Chance

(2)   Lack of physical or mental ability

(3)   Personality

(4)   Possibly, psychological problems.   Accidents may be a symptom of withdrawal from the work situation.

These last three categories do not seem to contribute a great deal to accidents in the process industries.

In some countries accidents may occur because employees' command of the written or spoken language used is poor or because their education and background make them reluctant to ask questions; the culture of the home and the factory may not be the same.[11]

4.4   Mind-sets

We have a problem. We think of a solution. We are then so busy congratulating ourselves that we fail to see that there may be a better solution, that some evidence points the other way, or that our solution has unwanted side-effects. This is known as a 'mind-set' or, if you prefer a more technical term, *Einstellung*.

De Bono writes, "Even for scientists there come a point in the gathering of evidence when conviction takes over and thereafter selects the evidence".[12]

Mind-sets are described by Raudsepp: [13]

> Most people when faced with a problem, tend to grab the first solution that occurs to them and rest content with it. Rare, indeed, is the individual who keeps trying to find other solutions to his problem. This is especially evident when a person feels under pressure. . . .
>
> And once a judgement is arrived at, we tend to persevere in it even when the evidence is overwhelming that we are wrong. Once an explanation is articulated, it is difficult to revise or drop it in the face of contradictory evidence. . . .
>
> Many interesting psychological experiments have demonstrated the fixating power of premature judgements. In one experiment, color slides of familiar objects, such as a fire hydrant, were projected upon a screen. People were asked to try to identify the objects while they were still out of focus. Gradually the focus was improved through several stages. The striking finding was this: If an individual wrongly identified an object while it was far out of focus, he frequently still could not identify it correctly when it was brought sufficiently into focus so that another person who had not seen the blurred vision could easily identify it. What this indicates is that considerably more effort and evidence is necessary to overcome an incorrect judgement, hypothesis or belief than it is to establish a correct one. A person who is in the habit of jumping to conclusions frequently closes his mind to new information, and limited awareness hampers creative solutions.

Mind-sets might have been discussed under "training", as the only way of avoiding them seems to be to make people aware of their existence and to discourage people from coming to premature conclusions. However, I have included them in this chapter as they are a feature of people's mental abilities which it is difficult to overcome and which we have to live with.

Here are some examples of mind-sets. Others were described in Sections 2.7.1.1 and 3.2 (Three Mile Island).

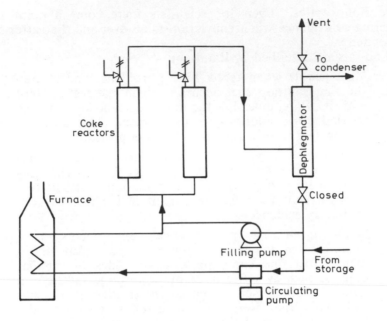

Figure 4.1   Simplified line diagram of coker.

4.4.1   An operator's mind-set

A good example is provided by an accident that occurred many years ago, on the coker shown, in a simplified form, in Figure 4.1. [14]

The accident occurred while the unit was being started up following a shut-down to empty the cokers of product, an operation that took place every few days. The normal procedure was to fill the plant with oil by opening the vent on the top of the dephlegmator and operating the low pressure filling pump and high pressure circulating pump in parallel. When oil came out of the vent, it was closed, the filling pump shut-down, circulation established and the furnace lit. This procedure, though primitive, was not unsafe as the oil had a high flash-point (32°C).

On the night of the accident the afternoon shift operator forgot to open the vent. When the night shift came on duty at 11.00 pm they found that the plant was filling with oil more slowly than usual and that the pump delivery pressure was higher than usual. As it was a very cold night they decided (about 2.00 am) that the low pumping rate and high pressure were due to an unusually high viscosity of the oil and they decided to light one burner in the furnace. Their diagnosis was not absurd, merely wrong. On earlier occasions lighting a burner had cured the same problem.

On this occasion it did not. The operators, however, were reluctant to consider that their theory might be wrong.

The filling pump got hot and had to be shut down. It was pumping against a rising pressure, which it could not overcome. The operators, however, ignored this clue and blamed the overheating of the pump on poor maintenance.

Finally, about 5.00 am the pump delivery pressure started to rise more rapidly. (Though far above normal for the filling stage, it was below the normal on-line operating pressure and so the relief valves did not lift.) The operators at last realised that their theory might be wrong. They decided to check that the vent valve was open. They found it shut. Before they could open it an explosion occurred killing one of the operators.

The cause of the explosion is interesting, though not related to the subject of this book. The dephlegmator acted as a giant slow-moving diesel engine, the rising level of oil compressing the air and oil vapour above it and raising their temperature until it exceeded the auto-ignition temperature of the oil vapour.

The incident is also interesting as an example of the slips discussed in Chapter 2. On the plant and a neighbouring one there had been 6000 successful start-ups before the explosion. The design and method of working made an error in the end almost inevitable but the error rate (1 in 6000) was very low, lower than anyone could reasonably expect. (1 in 1000 would be a typical figure.) However it may be that the vent valve had been left shut before but found to be so in time, and the incident not reported. If it had been reported, other shifts might have been more aware of the possible error.

4.4.2 A designer's mind-set

A new tank was designed for the storage of refrigerated liquid ethylene at low pressure (a gauge pressure of 0.8 psi or 0.05 bar). Heat leaking into the tank would vaporise some ethylene which was to be cooled and returned to the tank. When the refrigeration unit was shut-down — several days/year — the ethylene in the tank would be allowed to warm up a few degrees so that the relief valve lifted (at a gauge pressure of 1.5 psi or 0.1 bar) and the vapour would be discharged to a low stack (20 m high).

After construction had started, it was realised that on a still day the cold, heavy ethylene vapour would fall to ground level where it might be ignited. Various solutions were considered and turned down. The stack could not be turned into a flare stack, as the heat radiation at ground level would be too high. The stack could not be made taller as the base was too weak to carry the extra weight. Finally, someone had an idea: Put steam up the stack to warm up the ethylene so that it continued to rise and would not fall to the ground. This solution got everyone off the hook and was adopted.

When the plant was commissioned, condensate from the steam, running down the walls of the stack, met the cold ethylene gas and froze, completely blocking the 8 inches diameter stack. The tank was overpressured and split near the base. Fortunately the escaping ethylene did not ignite and the tank was emptied for repair without mishap. Afterwards a flare stack was constructed.

In retrospect, it seems obvious that ethylene vapour at -100°C might cause water to freeze, but the design team were so hypnotised by their solution that they were blind to its deficiencies. Use of a formal technique, such as a hazard and operability study,[15] for considering the consequences of changes would probably have helped.

This incident is interesting in another respect. For 11 hours before the split occurred the operators (on two shifts) were writing down on the record sheet readings above the relief valve set point; they steadied out at a gauge pressure of 2 psi (0.14 bar), the full scale deflection of the instrument. They did not realise the significance of the readings, and took no action. They did not even draw the foremen's attention to them and the foremen did not notice the high readings on their routine tours of the plant.

Obviously better training of the operators was needed, but in addition a simple change in the work situation would help: Readings above which action is required should be printed, preferably in red, on the record sheets.

4.4.3 Scholarly mind-sets

Mind-sets are not restricted to operators and designers. There are many examples of distinguished scholars who, having adopted a theory, continue to believe in it even though the evidence against it appears, to others, to be overwhelming. It may be interesting to describe two examples.

Some camelskin parchments were offered for sale in Jordon in 1966. The writing was similar to early Hebrew but not identical with it. A distinguished biblical scholar dated them to the 9th-7th centuries BC − 600 or 700 years earlier than the Dead Sea Scrolls − and suggested that the language was Philistine. If so, they would be the first documents yet discovered in that language.

Later, other scholars pointed out that the writing was a copy of a well-known early Hebrew inscription, but with the words in the wrong order. The documents were obviously fakes. The original identifier of the documents stuck to his opinion.

The parchments were dated by radio-carbon tests. These showed that they were modern. The scholar then said that the tests were meaningless; the parchments had been handled by so many people that they could have been contaminated.[16]

One classical scholar, convinced that the ancient Greeks could write nothing but great poetry, thought he had discovered the poetic metrical system for the poetry of the Linear B tablets — no mean achievement when one realizes that these tablets are administrative texts connected with the wool industry, the flax industry, copper-smithing and the manufacture of perfumes and unguents.[17]

4.4.4 A notorious mind-set

A frightening example of people's ability to see only the evidence that supports their view and ignore the rest is provided by the story of the witches of Salem, as told by Marion Starkey.[18]

References to Chapter 4

1. R G Sell, *Ergonomics versus accidents,* Ministry of Technology, Jan 1964.
2. *Petroleum Review,* April 1982, p 34.
3. *Petroleum Review,* Sept 1984, p 33.
4. M Gerard and J A B Hamilton, *Rails to Disaster,* Allen and Unwin, 1984, p 72, 69.
5. D M Hunns, *Terotechnica,* Vol 2, 1981, p 159.
6. J Love, *The Chemical Engineer,* No 403, May 1984, p 18.
7. D Hunter, *The Diseases of Occupations,* English Universities Press, 5th edition, 1975, p 1064.
8. A D Swain, *A Work Situation Approach to Improving Job Safety,* in J T Widner (editor), *Selected Readings in Safety,* Academy Press, Macon, Georgia, 1973, p 371.
9. H J Eysenck, *Fact and Fiction in Psychology,* Penguin Books, 1965, Chapter 6.
10. J M M Hill and E L Trist, *Industrial Accidents, Sickness and Other Absences,* Pamphlet No 4, Tavistock Publications, 1962.
11. A Foxcroft, *Safety Management* (South Africa), July 1984, p 29.
12. F de Bono, *An Atlas of Management Thinking,* Maurice Temple Smith, 1981, Penguin Books, 1983, p 129.
13. E Raudsepp, *Hydrocarbon Processing,* Vol 60, No 9, Sept 1981, p 29.
14. C H Vervalin (editor), *Fire Protection Manual for Hydrocarbon Processing Plant,* Gulf Publishing Co, 3rd edition, 1985, p 95.

15. T A Kletz, *Hazop and Hazan — Notes on the Identification and Assessment of Hazards,* Institution of Chemical Engineers, 1983.

16. *Biblical Archaeology Review,* Vol X, No 3, May/June 1984, p 66.

17. J D Muhly, *Biblical Archaeology Review,* Vol IX, No 5, Nov/Dec 1983, p 74.

18. M L Starkey, *The Devil in Massachusetts,* Knopf, 1949, Anchor Books, 1969. The witches of Salem are also the subject of Arthur Miller's play, *The Crucible.*

CHAPTER 5

ACCIDENTS DUE TO A LACK OF MOTIVATION

When I don't do it, I am lazy; when my boss
doesn't do it, he is too busy — Anon.

In this chapter we consider some accidents which occurred not because of forgetfulness, or lack of knowledge, or lack of ability, but because of a deliberate decision not to do something. They may be conveniently divided into management decisions and operator decisions, depending on the level at which the decision was made. If operators cut corners it may be because they are not convinced that the procedure is necessary, in which case the accident is really due to a lack of training (see Chapter 2). It may be, however, that they cut corners because all men carrying out a routine task become careless after a time. Managers should carry out formal audits or inspections, or just keep their eyes open, in order to see that the proper procedures are being followed.

The division into two categories is not, of course, absolute. Junior managers may need training themselves, if they are not convinced of the need for safety procedures, and they may also cut corners if *their* bosses do not check up from time to time.

As with the forgetfulness incidents discussed in Chapter 2, whenever possible we should look for engineering solutions — designs which are not dependent on the operator carrying out a routine task correctly — but very often we have to adopt a software solution, that is, we have to persuade or cajole people to follow good practice.

5.1   Accidents due to deliberate management decisions

Accidents of this type have been particularly emphasised by W B Howard, one of the best-known US loss prevention engineers, particularly in a paper sub-titled "We aint farmin' as good as we know how".[1] He tells the story of a young graduate in agriculture who tried to tell an old farmer how to improve his methods. The farmer listened for a while and then said, "Listen son, we aint farmin' now as good as we know how".

Howard describes several accidents which occurred because managers were not 'farming' as well as they could — and should — have done. For example:

An explosion occurred in a hold tank in which reaction product had been kept for several hours. It was then found that:

— The pressure had been rising for several hours before the explosion but the operator on duty that day had had no training for the operation and did not realise the significance of the pressure rise.

— No tests had been made to see what happened when the reaction product was kept for many hours. Tests had been repeatedly postponed because everyone was "too busy" running the unit.

The accident was not the result of equipment failure or human error as usually thought of, that is, forgetfulness (see Chapter 2) but rather the result of conscious decisions to postpone testing and to put an inexperienced operator in charge.

Howard also describes a dust explosion and fire which occurred because the operating management had by-passed all the trips and alarms in the plant, in order to increase production by 5%. Before you blame the operating management, remember they were not working in a vacuum. They acted as they did because they sensed that their bosses put output above safety. In these matters official statements of policy count for little. Little things count for more. When senior managers visited the plant, did they ask about safety or just about output?

Many years ago, when I was employed on operations, not safety, my works manager changed. After a few months, someone told the new works manager that the employees believed him to be less interested in safety than his predecessor. He was genuinely shocked. "Whatever have I said", he asked, "to give that impression? Please assure everyone that it is wrong".

It was not what he had said that created the impression but what he had not said. Both managers spent a good deal of their time out on the plant, and were good at talking to operators, supervisors and junior managers. The first works manager, unlike the second, frequently brought safety matters into the conversation.

Another example of a deliberate management decision was described in Hansard.[2] After a disgruntled employee had blown the whistle, a factory inspector visited a site where there were six tanks containing liquefied flammable gas. Each tank was fitted with a high level alarm and a high level trip. The factory inspector found that five of the alarms were not working, that no-one knew how long they had been out-of-order and that the trips were never tested, could not be tested and were of an unreliable design.

In this case the top management of the organisation concerned was committed to safety, as they employed a large team of safety specialists in their head office, estimating accident probabilities. (Their calculations were useless as the assumptions on which they were based — that trips and alarms would be tested — were not correct. They would have been better employed out on the plant testing the alarms.) However the top management had failed to get the message across to the local management who took a deliberate decision not to test their trips and alarms.

When managers make a deliberate decision to stop an activity, they usually let it quietly lapse and do not draw attention to the change. However one report frankly stated, "The data collection system . . . . . was run for a period of about 3 years . . . . . The system is now no longer running, not because it was unsuccessful, but because the management emphasis on the works has changed. The emphasis is now on the reduction of unnecessary work, not on the reduction of breakdowns."[3]

5.2 Accidents due to operator neglect

These are also deliberate decisions, but easily made. When operators have to carry out routine tasks, and an accident is unlikely if the task, or part of it, is omitted or shortened, the temptation to take short cuts is great. It is done once, perhaps because the operator is exceptionally busy or tired, then a second time and soon becomes routine. Some examples follow.

5.2.1 Preparation for maintenance

Many supervisors find permit-to-work procedures tedious. Their job, they feel, is to run the plant, not fill in forms. There is a temptation, for a quick job, not to bother. The fitter is experienced, has done the job before, so let's just ask him to fix the pump again.

They do so, and nothing goes wrong, so they do so again. Ultimately the fitter dismantles the wrong pump, or the right pump at the wrong time, and there is an accident.

Or the fitter does not bother to wear the proper protective clothing requested on the permit-to-work. No-one says anything, to avoid unpleasantness; ultimately the fitter is injured.

To prevent these accidents a two-pronged approach is necessary:

(1) We should try to convince people why a permit-to-work is necessary by describing accidents that have occurred because there was no adequate system or the system was not followed. Discussions are better than lectures or reports and the Institution of Chemical Engineers' Hazard Workshop Module[4] can be used for this purpose.

(2) Managers should check from time to time that the correct procedures are being followed. A friendly word the first time someone takes a short cut is more effective than punishing them after an accident has occurred.

If an accident is the result of taking a short cut, it is unlikely that it occurred the first time the short cut was taken. It is more likely that short cutting has been going on for weeks or months. A good manager would have spotted it and stopped it. If he does not, then when the accident occurs he shares the responsibility for it, legally and morally, even though he is not on the site at the time.

A manager is not, of course, expected to stand over his team at all times. But he should carry out periodic inspections to check that procedures are being followed and he should not turn a blind eye when he sees unsafe practices in use. (See Section 11.2.)

The first step down the road to a serious accident occurs when a manager turns a blind eye to a missing blind.

5.2.2   An incident on a hydrogen/oxygen plant

An explosion occurred on a plant making hydrogen and oxygen by the electrolysis of water. As a result of corrosion some of the hydrogen had entered the oxygen stream.

Both streams were supposed to be analysed every hour. After the explosion, factory inspectors went through old record sheets and found that when conditions changed the analytical results on the sheets changed at once, although it would take an hour for a change to occur on the plant. It was obvious that the analyses were not being done.

The management had not noticed this and had failed to impress on the operators the importance of regular analyses and the results of not detecting hydrogen in the oxygen. (See also Section 4.2.2.)

Engineering solutions, when possible, are usually better than reliance on manual tests and operator action and the official report recommended automatic monitoring and shut-down.[5] (But see Section 6.5.2.)

5.2.3   An example from the railways

The railways — or some of them — realised early on that when possible designers should make it difficult or impossible for people to by-pass safety equipment. Engine drivers were tempted to tamper with the relief valves on early locomotives. As early as 1829, before their line was complete, the Liverpool and Manchester Railway laid down the following specification for their engines:

> "There must be two safety valves, one of which must be completely out of reach or control of the engine man."
> (Figure 12.13).

Incidentally, contrary to popular ·belief, very few locomotive boilers blew up because the driver tampered with the relief valve — a view encouraged by locomotive superintendents. Most explosions were due to corrosion — preventable by better inspection or better design, — poor-maintenance or delaying maintenance to keep engines on the road. Many of the famous locomotive designers were abysmal maintenance engineers.[6]

5.2.4 Taking a chance

A man went up a ladder onto a walkway and then climbed over the handrails onto a fragile roof which was clearly labelled. He tried to stand only on the cross girders but slipped off onto the roof and fell through into the room below.

The accident was put down to 'human failing', the failure of the injured man to follow the rules.

When the site of the accident was visited however it was found that the foot of the ladder leading to the roof was surrounded by junk, as shown in Figure 5.1. It shows that too little attention was paid to safety in the plant and that in 'taking a chance' the injured man was following the example set by his bosses. They were, to some extent, responsible for the accident.

This incident also show the importance, in investigating an accident, of visiting the scene and not just relying on reports.

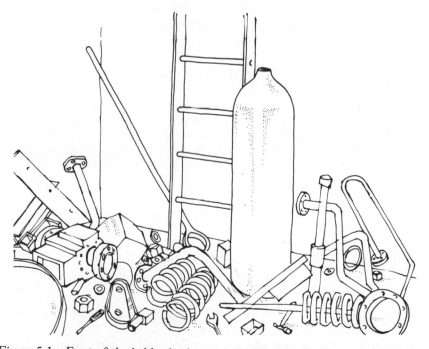

Figure 5.1 Foot of the ladder leading to a fragile roof. Did this scene encourage a man to take a chance?

5.2.5 Motivational Appeals

In case anyone is tempted by them, perhaps I should say that little or no improvement will result from generalised exhortations to people to work safely, follow the rules, be responsible or otherwise avoid sin. (See also Section 3.6.3.)

Swain writes:

> "Motivational appeals have a temporary effect because man adapts. He learns to tune out stimuli which are noise, in the sense of conveying no information. Safety campaigns which provide no useful, specific information fall into the "noise" category. They tend to be tuned out."

> And ". . . a search of the literature of industrial psychology has failed to show a single controlled experiment on the real effectiveness of safety motivation campaigns".[7]

If people are not following the rules or working safely we should:

(1) First check that it is possible for them to do so or if any changes in plant design or method of working would make it easier.

(2) Explain why it is necessary to follow the rules and back this up by describing accidents which occurred because the rules were not followed. If you have failed to convince the workforce as a whole that an action or procedure is necessary, then there is little you can do to enforce it.

(3) Check up to see that the rules are being followed and point out to people when they are not. Most people will respond to this treatment. Most people do what the boss wants; it makes for an easier life. Only on rare occasions will it be necessary to 'get tough', and then only with an occasional individual.

References to Chapter 5

1. H B Howard, *Plant/Operations Progress,* Vol 3, No 3, July 1984, p 147.

2. *Hansard,* 8 May 1980.

3. Systems Reliability Service, *Report of Annual General Meeting,* 1974, p 74.

4. *Hazard Workshop Module No 004, Preparation for Maintenance,* Institution of Chemical Engineers.

5.   Health and Safety Executive, *The Explosion at Laporte Industries Limited on 5 April 1975*, HMSO, 1976.

6.   C H Hewison, *Locomotive Boiler Explosions*, David and Charles, 1983.

7.   A D Swain, *A Work Situation Approach to Improving Job Safety*, in J T Widner (editor), *Selected Readings in Safety*, Academy Press, Macon, Georgia, 1973, p 371.

CHAPTER 6

THE PROBABILITY OF HUMAN ERROR

". . . . I tried to talk my staff into doing it as they, at least, work for me. And if they put a number in I didn't like, I could jiggle it around" — W Akerm, (giving evidence at an inquiry), quoted in H C Kunreuther and J Linnerooth, *Risk Analysis and Decision Processes,* Springer-Verlag, 1983

In this chapter I list some estimates that have been made of the probability of error. However several limitations of these data should be borne in mind.

First, the figures are estimates of the probability that someone will, for example, have a moment's forgetfulness and forget to close a valve or have a moment's aberration and close the wrong valve, press the wrong button, make a mistake in arithmetic and so on, errors of the type described in Chapter 2. They are not estimates of the probability of error due to poor training or instructions (Chapter 3), lack of physical or mental ability (Chapter 4) or lack of motivation (Chapter 5). There is no way of estimating the probabilities of such errors.

Second, the figures are mainly estimates by experienced managers and are not based on a large number of observations. Where operations such as soldering electric components are concerned, observed figures based on a large sample are available but this is not the case for typical process industry tasks.

Because so much judgment is involved, it is tempting for those who wish to do so to try to 'jiggle' the figures to get the answer they want. (The author of the quotation at the head of this chapter was not speaking particularly of figures on human error but there is no branch of hazard analysis to which his remarks are more applicable.) Anyone who uses estimates of human reliability outside the usual ranges (see later) should be expected to justify them.

Third, the actual error rates depend greatly on the degree of stress and distraction. The figures quoted are for typical process industry conditions.

Fourth, every man is different and no-one can estimate the probability that he will make a mistake. All we can estimate is the probability that, if a large number of men were in a similar situation, mistakes would be made.

6.1 Why do we need to know human error rates?

In brief, because men are part of the total protective system and we cannot estimate the reliability of the total protective system unless we know the reliability of each part of it, including the man.

Consider the situation shown in Figure 6.1. When the alarm sounds and flashes in the control room the operator is expected to go outside, select the right valve out of many and close it within, say, 10 minutes. We can estimate fairly accurately the reliability of the alarm and if we think it is too low it is easy to improve it. We can estimate roughly the reliability of the valve — the probability that the operator will be able to turn the handle and that the flow will actually stop — and if we think it is too low we can use a better quality valve or two in series. But what about the man in the middle? Will he always do what he is expected to do?

In my time in industry, people's opinions varied from one extreme to another. At times people, particularly design engineers, said that it was reasonable to expect him to always close the right valve. If he did not he should be reprimanded. I hope that the incidents described in Chapter 2 will have persuaded readers of the impracticality of this approach.

At other times people, particularly managers responsible for production, have said that it is well-known that men are unreliable. We should therefore install fully automatic equipment, so that the valve is closed automatically when the alarm condition is reached.

Both these attitudes are unscientific. We should not say "The operator always will . . ." or "The operator never will . . ." but ask "What is the probability that the operator will . . ." Having agreed a figure we can feed it into our calculations. If the consequent failure rate is too high, we can consider a fully automatic system. If it is acceptable, we can continue with the present system.

6.2 Human error rates — a simple example

The following are some figures I have used for failure rates in the situation described, that is, the probability that a typical operator will fail to close the right valve in the required time, say 10 minutes. The estimates are conservative figures intended for design, rather than hazard assessment, purposes.

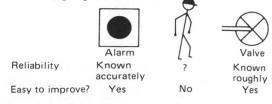

	Alarm		Valve
Reliability	Known accurately	?	Known roughly
Easy to improve?	Yes	No	Yes

Figure 6.1   Reliabilities in a man/machine system.

1. *When failure to act correctly will result in a serious accident such as a fire or explosion* — 1 (1 in 1)

The operator's failure rate will not really be as high as this, but it will be very high (more than 1 in 2) and we should assume 1 in 1 for design purposes.

2. *In a busy control room* — 0.1 (1 in 10)

This may seem high but before the operator can respond to the alarm another alarm may sound, the telephone may ring, a fitter may demand a permit-to-work and another operator may report a fault outside. Also, on many plants the control room operator cannot leave the control room and has to contact an outside man, by radio or loudspeaker, and ask him to close the valve. This provides opportunities for misunderstanding.

3. *In a quiet control room* — 0.01 (1 in 100)

Busy control rooms are the more common, but control rooms in storage areas are usually quieter places with less stress and distraction. In some cases a figure between 1 in 10 and 1 in 100 can be chosen.

4. *If the valve is immediately below the alarm* — 0.001 (1 in 1000)

The operator's failure rate will be very low but an occasional error may still occur. A failure rate of 1 in 1000 is about the lowest that should be assumed for any process operation. For example, failure to open a valve during a routine operation, as described in Section 2.2.

6.3   A more complex example

Figure 6.2 shows the fault tree for loss of level in a distillation column followed by breakthrough of vapour at high pressure into the downstream storage tank. A line diagram and further details

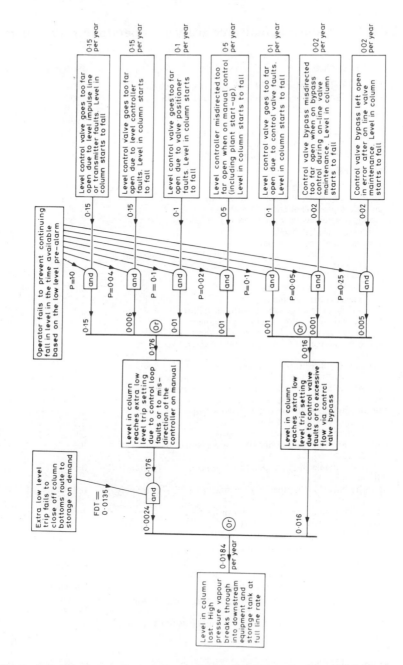

Figure 6.2 Fault tree for vapour breakthrough from high pressure distillation column into downstream equipment due to loss of level in distillation column base (orginal design).

73

TABLE 6.1   Industrial fault tree analysis: operator error estimates.

Crystalliser Plant		Probability
Operator fails to observe level indicator or take action		0.04
Operator fails to observe level alarm or take action		0.03
		Frequency events/y
Manual isolation valve wrongly closed (p)		0.05 and
Control valve fails open or misdirected open		0.5
Control valve fails shut or misdirected shut (1)		0.5

Propane Pipeline		
Operator fails to take action:	Time available	Probability
to isolate pipeline at planned shutdown		0.001
to isolate pipeline at emergency shutdown		0.005
opposite spurious tank blowdown given alarms and flare header signals	30 min	0.002
opposite tank low level alarm		0.01
opposite tank level high given alarm with	5-10 min	
a) controller misdirected or bypassed when on manual		0.025
b) level measurement failure		0.05
c) level controller failure		0.05
d) control valve or valve positioner failure		0.1
opposite slowly developing blockage on heat exchanger revealed as heat transfer limitation		0.04
opposite pipeline fluid low temperature given alarm	5 min	0.05
opposite level loss in tank supplying heat transfer medium pump given no measurement (p)	5 min	0.2
opposite tank blowdown without prior pipeline isolation given alarms which operator would not regard as significant and pipework icing	30 min	
a) Emergency blowdown		0.2
b) Planned blowdown		0.6
opposite pipeline fluid temperature low given alarm		0.4
opposite pipeline fluid temperature low given alarm	limited	0.8
opposite backflow in pipeline given alarm	extremely short	0.8
opposite temperature low at outlet of heat exchanger given failure of measuring instrument common to control loop and alarm		1
Misvalving in changeover of two-pump set (standby pump left valved open, working pump left valved in)		0.0025/ changeover
Pump in single or double pump operation stopped manually without isolating pipeline		0.01/ shutdown
LP steam supply failure by fracture, blockage or isolation error (p)		Frequency 0.1/y
Misdirection of controller when on manual (assumed small proportion of time)		1/y

Notes:

1 = literature value
p = plant value
Other values are assumptions

TABLE 6.2   TESEO: error probability parameters

Type of activity

	K_1
Simple, routine	0.001
Requiring attention, routine	0.01
Not routine	0.1

Temporary stress factor for routine activities

Time available (s)	K_2
2	10
10	1
20	0.5

Temporary stress factor for non-routine activities

Time available (s)	K_2
3	10
30	1
45	0.3
60	0.1

Operator qualities

	K_3
Carefully selected, expert, well-trained	0.5
Average knowledge and training	1
Little knowledge, poorly trained	3

Activity anxiety factor

	K_4
Situation of grave emergency	3
Situation of potential emergency	2
Normal situation	1

Activity ergonomic factor

	K_5
Excellent microclimate, excellent interface with plant	0.7
Good microclimate, good interface with plant	1
Discrete microclimate, discrete interface with plant	3
Discrete microclimate, poor interface with plant	7
Worst microclimate, poor interface with plant	10

including a detailed quantitative hazard analysis are given in Reference 1.

The right-hand column of the tree shows the "bottom events", the initiating events that can lead to loss of level. An alarm warns the operator that the level is falling. The tree shows the probabilities that the operator will fail to prevent a continuing fall in level in the time available.

For the top branch this probability (P) is 1 because the alarm and the level indicator in the control room would also be in a failed condition. The operator could not know that the level was falling.

For the other branches values of P between 0.02 and 0.25 have been estimated. They include an allowance (P = 0.005) for the coincidental failure of the alarm.

For the fourth branch P has been assumed to be 0.02 (2% or 1 in 50) as the operator is normally in close attendance when a controller is on manual and correction is possible from the control room.

For the second branch P has been assumed to be 0.04 (4% or 1 in 25) as conditions are more unexpected and the operator is more likely to be busy elsewhere.

For the third and fifth branches P = 0.1 (10% or 1 in 10) as the operator has to ask an outside man to adjust a manual valve. The control room operator may delay making the request in order to make sure that the level really is falling.

For the sixth branch P = 0.05 (5% or 1 in 20). The outside operator has to be contacted but he should be near the valve and expecting to be contacted.

Finally, for the last branch P = 0.25 (25% or 1 in 4). The fault is unusual and the outside man may overlook it.

In the study of which Figure 6.2 formed a part, the estimates of operator reliability were agreed between an experienced hazard analyst and the commissioning manager. This reduces the chance that the figures will be 'jiggled' (see quotation at the head of this chapter) to get a result that the designer, or anyone else, would like. The figures used apply to the particular plant and problem and should not be applied indiscriminately to other problems. For example, if loudspeakers are inaudible in part of the plant or if several outside operators can respond to a request, so that each leaves it to the others (see Section 3.3.6), the probabilities that they will fail will be high.

In another similar study Lawley[2] used the figures set out in Table 6.1 in a form used by Lees.[3]

6.4 Other estimates of human error rates

Bello and Columbori[4] have devised a method known as TESEO (Technica Empirica Stima Errori Operati). The probability of error is assumed to be the product of five factors – K_1–K_5 – which are defined and given values in Table 6.2

Swain[5] has developed a method known as THERP (Technique for Human Error Rate Prediction). The task is broken down into individual steps and the probability of error estimated for each, taking into account:

— the likelihood of detection

— the probability of recovery

— the consequence of the error (if uncorrected)

— a series of 'performance shaping factors' such as temperature, hours worked, complexity, availability of tools, fatigue, monotony, and group identification (about 70 in total).

Table 6.3 is a widely quoted list of error probabilities taken from the US Atomic Energy Commission Reactor Safety Study (the Rasmussen Report).[6]

A study of batch chemical operations produced the following results:

	Probability per batch	
Ingredient omitted	2.3×10^{-4}	(1 in 4375 batches)
Ingredient undercharged	1.1×10^{-4}	(1 in 8750 batches)
Ingredient overcharged	2.6×10^{-4}	(1 in 3900 batches)
Wrong ingredient added	2.3×10^{-4}	(1 in 4375 batches)
Total errors	8.3×10^{-4}	(1 in 1200 batches)

These error rates seem rather low. However they do not include:

— Errors which were not reported — estimated at 50% of the total

— Errors which were corrected — estimated at 80% of the total

— Errors which were too small to matter — estimated at 80% of the total

The total error rate will then be 4×10^{-2} per batch or 1 in 25 batches.

There are on average four charging operations per batch so the error rate becomes 10^{-2} per operation (1 in 100 operations) which is in line with other estimates.

6.5 Two more simple examples

6.5.1 Starting a spare pump

As an example of the uses of some of these figures, let us consider another simple process operation: starting up a spare pump after the running pump has tripped out.

TABLE 6.3 General estimates of error probability used in US Atomic Energy Commission reactor safety study.

Estimated error probability	Activity
10^{-4}	Selection of a key-operated switch rather than a non-key switch (this value does not include the error of decision where the operator misinterprets situation and believes key switch is correct choice)
10^{-3}	Selection of a switch (or pair of switches) dissimilar in shape or location to the desired switch (or pair of switches), assuming no decision error. For example, operator actuates large-handled switch rather than small switch
3×10^{-3}	General human error of commission. e.g. misreading label and therefore selecting wrong switch
10^{-2}	General human error of omission where there is no display in the control room of the status of the item omitted, e.g. failure to return manually operated test valve to proper configuration after maintenance
3×10^{-3}	Errors of omission, where the items being omitted are embedded in a procedure rather than at the end as above
3×10^{-2}	Simple arithmetic errors with self-checking but without repeating the calculation by re-doing it on another piece of paper
$1/x$	Given that an operator is reaching for an incorrect switch (or pair of switches), he selects a particular similar appearing switch (or pair of switches), where x=the number of incorrect switches (or pair of switches) adjacent to the desired switch (or pair of switches). The $1/x$ applies up to 5 or 6 items. After that point the error rate would be lower because the operator would take more time to search. With up to 5 or 6 items he does not expect to be wrong and therefore is more likely to do less deliberate searching
10^{-1}	Given that an operator is reaching for a wrong motor operated valve (MOV) switch (or pair of switches), he fails to note from the indicator lamps that the MOV(s) is (are) already in the desired state and merely changes the status of the MOV(s) without recognizing he had selected the wrong switch(es)
~ 1.0	Same as above, except that the state(s) of the incorrect switch(es) is (are) *not* the desired state
~ 1.0	If an operator fails to operate correctly one of two closely coupled valves or switches in a procedural step, he also fails to correctly operate the other valve.
10^{-1}	Monitor or inspector fails to recognize initial error by operator. *Note:* With continuing feedback of the error on the annunciator panel, this high error rate would not apply
10^{-1}	Personnel on different work shift fail to check condition of hardware unless required by checklist or written directive
5×10^{-1}	Monitor fails to detect undesired position of valves, etc., during general walk-around inspections, assuming no checklist is used
0.2-0.3	General error rate given very high stress levels where dangerous activities are occurring rapidly
$2^{(n-1)}x$	Give severe time stress, as in trying to compensate for an error made in an emergency situation, the initial error rate, x, for an activity doubles for each attempt, n, after a previous incorrect attempt, until the limiting condition of an error rate of 1.0 is reached or until time runs out. This limiting condition corresponds to an individual's becoming completely disorganized or ineffective
~ 1.0	Operator fails to act correctly in first 60 seconds after the onset of an extremely high stress condition, e.g. a large LOCA
9×10^{-1}	Operator fails to act correctly after the first 5 minutes after the onset of an extremely high stress condition
10^{-1}	Operator fails to act correctly after the first 30 minutes in an extreme stress condition
10^{-2}	Operator fails to act correctly after the first several hours in a high stress condition
x	After 7 days after a large LOCA, there is a complete recovery to the normal error rate, x, for any task

Notes: (1) Modifications of these underlying (basic) probabilities were made on the basis of individual factors pertaining to the tasks evaluated.

(2) Unless otherwise indicated, estimates or error rates assume no undue time pressures or stresses related to accidents.

78

Many analysts would use the simple approach of Section 6.2 and assume that the job will be done correctly 99 times out of 100 in a normal, low stress situation, rather less often — perhaps 9 times out of 10 — if the stress is high — say the operator knows that the plant will shut down in 5 minutes if the spare pump is not started up correctly.

Let us see if a more analytical approach is helpful.

The task can be split into a number of steps:

1. Walk to pump,

2. Close delivery valve of failed pump,

3. Close suction valve of failed pump,

4. Open suction valve of spare pump,

5. Press start button,

6. Open delivery valve of spare pump.

It does not matter if the operator forgets to carry out steps 2 and 3, so there are four steps which have to be carried out correctly.

Step 1 is included as perhaps one of the commonest sources of error is failing to carry out this step, that is, the operator forgets the whole job because he has other things on his mind, or goes to the wrong pump.

From Table 6.3, the lines which seem most applicable, in a low stress situation, are lines 3 and 5.

Estimated error probability	Activity
3×10^{-3}	General human error of commission, eg, mis-reading label and therefore selecting wrong switches
3×10^{-3}	Errors of omission, where the items being omitted are embedded in a procedure

There are 4 critical steps, so the total probability of error is
$$12 \times 10^{-3} = 0.012 \ (1 \text{ in } 80)$$

Table 6.3 is not much help to us in a condition of moderate stress, though it does consider very high stress situations (last 5 lines), so let us try applying Table 6.2.

Type of activity: Requiring attention, routine $K_1 = 0.01$

79

Stress factor:	More than 20 secs available	K_2	= 0.5
Operator qualities:	Average knowledge and training	K_3	= 1
Activity anxiety factor:	Potential emergency	K_4	= 2
Activity ergonomic factor:	Good microclimate, good interface with plant	K_5	= 1

Probability of error = $K_1 \ K_2 \ K_3 \ K_4 \ K_5$
          = 0.01

If these figures are assumed to apply to each step , the total probability of error is 0.04 (1 in 25).

However if we assume each step is "simple", rather than one "requiring attention", the error rate is 10 times lower. This illustrates the limitation of these techniques, when successive figures in a table differ by an order of magnitude and shows how easy it is for an unscrupulous person to 'jiggle' the figures to get the answer he wants. The techniques are perhaps most valuable in estimating relative error probabilities rather than absolute values.

It is interesting to compare these estimates with the reliability of an automatic start mechanism. A typical failure rate will be 0.25/year and if the mechanism is tested every 4 weeks, its fractional dead time (probability of failure on demand) will be:

$$\tfrac{1}{2} \times 0.25 \times \tfrac{4}{52} \quad = 0.01 \text{ or } 1 \text{ in } 100$$

None of our estimates of human reliability is lower than this, thus confirming the instinctive feel of most engineers that in this situation an automatic system is more reliable and should be installed if failure to start the pump has serious consequences, but it is hardly justified in the normal situation.

6.5.2 Filling a tank

Suppose a tank is filled once/day and the operator watches the level and closes a valve when it is full. The operation is a very simple one, with little to distract the operator who is out on the plant giving the job his full attention. Most analysts would estimate a failure rate of 1 in 1000 occasions or about once in 3 years. In practice, men have been known to operate such systems for 5 years without incident.

This is confirmed by Table 6.2 which gives:

$$K_1 = 0.001$$
$$K_2 = 0.5$$
$$K_3 = 1$$
$$K_4 = 1$$
$$K_5 = 1$$

Failure rate = 0.5×10^{-3} or 1 in 2000 occasions (6 years).

An automatic system would have a failure rate of about 0.5/year and as it is used every day testing is irrelevant and the hazard rate (the rate at which the tank is overfilled) is the same as the failure rate, about once every 2 years. The automatic equipment is therefore less reliable than an operator.

Replacing the operator by automatic equipment will *increase* the number of spillages unless we duplicate the automatic equipment or use Rolls Royce quality equipment (if available). (For a more detailed treatment of this problem see Reference 7.)

Note that if we replace an operator by automatic equipment we do not, as is often thought, eliminate the human element. We may remove our dependence on the operator but we are now dependent on the men who design, construct, install, test and maintain the automatic equipment. We are merely dependent on different men. It may be right to make the change (it was in one of the cases we have considered, not the other) as these men usually work under conditions of lower stress than the operator but do not let us kid ourselves that we have removed our dependence on men.

6.5.3 More opportunities – more errors

In considering errors such as those made in starting a pump, filling a tank, etc, do not forget that the actual *number* of errors made by an operator, as distinct from the *probability* of errors, depends on the number of times he is expected to start a pump, fill a tank, etc.

I once worked in a works that consisted of two sections: large continuous plants and small batch plants. The foremen and operators on the batch plants had a poor reputation as a gang of incompetents who were always making mistakes: overfilling tanks, putting material in the wrong tank, etc. Some of the best men on the continuous plants were transferred to the batch plants but with little effect. Errors *rates* on the batch plants were actually lower than on the continuous plants but there were more opportunities for error: pumps were started up, tanks filled, etc many more times per day.

6.6   Button pressing

The American Institute for Research have published a series of papers on the reliability of simple operations such as those used in operating electronic equipment. [8]   The application and limitations of their data book can be illustrated by applying it to one of the button-pressing operations described in Section 2.4.1, the operation of a beverage vending machine.

The push buttons are considered under several headings:

The first is size. The probability that the correct button will be pressed depends on the size as shown below:-

Miniature	0.9995
$\frac{1}{2}$ inch	0.9999
more than $\frac{1}{2}$ inch	0.9999 ←

The pushbuttons on the beverage machines are $1\frac{1}{2}$ inch by $\frac{5}{8}$ inch so I have put an arrow against the last item.
Next we consider the number of pushbuttons in the group.
Single column or row:

1-5	0.9997
4-10	0.9995 ←
11-25	0.9990

The next item to be considered is the number of pushbuttons to be pushed

2	0.9995
4	0.9991
8	0.9965

On a beverage machine only one button has to be pushed so I assumed that the probability of success is 0.9998.

The fourth item is the distance between the edges of the buttons

$\frac{1}{8}'' - \frac{1}{4}''$	0.9985 ←
$\frac{3}{8}'' - \frac{1}{2}''$	0.9993
$\frac{1}{2}''$ up	0.9998

The fifth item is whether or not the button stays down when pressed

Yes	0.9998 ←
No	0.9993

The final item is clarity of the labelling.

At least 2 indications of control positioning	0.9999
Single, but clear and concise indication of control positioning	0.9996 ←
A potentially ambiguous indication of control positioning	0.9991

The total probability of success is obtained by multiplying these six figures together:-

$$\text{Reliability} = 0.9999 \times 0.9995 \times 0.9998 \times 0.9985 \times 0.9998 \times 0.9996$$
$$= 0.9971$$

ie., three errors can be expected in every 1000 operations.

I actually got the wrong drink about once in 50 times (that is, 20 in 1000 times) that I used the machines. Perhaps I am seven times more careless than the average man, or inadequately trained or physically or mentally incapable. It is more likely that the difference is due to the fact that the method of calculation makes no allowance for stress or distraction and that the small amount present is sufficient to increase my error rate seven times. (The machines are in the corridor, so I may talk to the people who pass, or I may be thinking about the work I have just left.)

6.7   Non-process operations

As already stated, for many assembly line and similar operations error rates are available based not on judgment but on a large data base. They refer to normal, not high stress, situations. Some examples follow. Remember that many errors can be retrieved and that not all errors matter (or cause degradation of mission fulfilment, to use the jargon used by many workers in this field).

A D Swain[9] quotes the following figures for operations involved in the assembly of electronic equipment:

	Error Rate per Operation
Excess solder	0.0005
Insufficient solder	0.002
Two wires reversed	0.006
Capacitor wired backwards	0.001

He quotes the following error rates for inspectors measuring the sizes of holes in a metal plate:

Errors in addition, subtraction and division	0.013
Algebraic sign errors	0.025
Measurement reading errors	0.008
Copying errors per 6-digit number	0.004

These should not, of course, be used uncritically in other contexts.

6.8   Train driver errors

Section 2.7.1.2 discussed errors by train drivers who passed signals at danger so it may be interesting to quote an estimate of their error rates.[10]

In 1972, in the Eastern Region of British Rail, 91 signals were passed at danger (information from British Rail). 80-90% of these, say 83, were due to drivers' errors. The rest were due to other causes such as signals being suddenly altered as the train approached. It is estimated that many incidents are not reported and that the true total would be 2-2.5 times this figure, say 185. The total mileage covered in the Eastern Region in 1972 was 72.2×10^6 miles (information from British Rail). The spacing of signals is about 1 every 1,200 yards (0.7 mile) on the main lines; rather less frequent on branch lines. Say 0.75 mile on average. Therefore, in 1972, $72.2 \times 10^6/0.75 = 96 \times 10^6$ signals were passed. Therefore, the chance that a signal will be passed at danger is 1 in $96 \times 10^6/185 = 1$ in 5×10^5 signals approached (*not* signals at danger passed). If we assume that between 1 in 10 and 1 in 100 (say 1 in 35) signals approached is at danger then a signal will be passed at danger once every $5 \times 10^5/35 \approx 10^4$ occasions that a signal at danger is approached. This is at the lower end of the range quoted for human reliability.

Thanks are due to Mr J M Boyes and Mr P W B Semmens who supplied the estimates used in this calculation.

6.9   Some pitfalls in using data on human reliability

6.9.1   Checking may not increase reliability

If a man knows he is being checked, he works less reliably. If the error rate of a single operator is 1 in 100, the error rate of an operator plus a checker is certainly greater than 1 in 10 000 and may even be greater than 1 in 100, that is, the addition of the checker say actually increases the overall error rate.

If two men work together as a team, error rates come down. Ref 11 describes the calibration of an instrument in which one man writes down the figures on a check list while the other man calls them out. The two men then change over and repeat the calibration. The probability of error was put at 10^{-5}.

Requiring a man to sign a statement that he has completed a task produces very little increase in reliability as it soon becomes a perfunctory activity.[11]

6.9.2   Increasing the number of alarms does not increase reliability proportionately

Suppose an operator ignores an alarm on 1 in 100 of the occasions on which it sounds. Installing another alarm (at a slightly different setting or on a different parameter) will not reduce the failure rate to 1 in 10 000. If the operator is in a state in which he ignores the first alarm, then there is a more than average chance that he will ignore the second. (In one plant there were five alarms in series. The operator was assumed to ignore each one on one occasion in ten, the whole lot on one occasion in 100 000!).

6.9.3   If an operator ignores a reading he may ignore the alarm

Suppose an operator fails to notice a high reading on one occasion in 100 – it is an important reading and he has been trained to pay attention to it. Suppose that he ignores the alarm on one occasion in 100. Then we cannot assume that he will ignore the reading and the alarm on one occasion in 10 000. On the occasion on which he ignores the reading the chance that he will ignore the alarm is greater than average.

While this book was in the press a review appeared of attempts to validate human reliability assessment techniques (J C Williams, *Reliability Engineering,* Vol 11, No 3, 1985, p 149). The author concludes, "Even the most comprehensive study of assessment techniques cannot tell us whether any of the methods it evaluated are worth much further consideration. Perhaps the only real message to emerge is that Absolute Probability Judgment (APJ) was the least worse of the methods evaluated".

APJ is another term for experienced judgment or, if you prefer, guessing.

References to Chapter 6

1. T A Kletz and H G Lawley, Chapter 2.1 of A E Green (editor), *High Risk Safety Technology,* Wiley, 1982.

2. H G Lawley, *Reliability Engineering,* Vol 1, 1980, p 89.

3. F P Lees, *The Assessment of Human Reliability in Process Control,* Conference on Human Reliability in the Process Control Centre: Institution of Chemical Engineers, Manchester, 1983.

4. G C Bello and V Columbori, *Reliability Engineering,* Vol 1, No 1, July-Sept 1980, p 3.

5. A D Swain and H E Gutterman, *Handbook of Human Reliability Analysis with Emphasis on Nuclear Power Plant Applications,* Report No NUREG/CR-1278, Sandia Laboratories, Albuquerque, New Mexico, 1980.

6. Atomic Energy Commission, *Reactor Safety Study: An Assessment of Accident Risk in US Commercial Nuclear Power Plant,* Report No WASH 1400, Atomic Energy Commission, Washington DC, 1975.

7. T A Kletz, *Hazop and Hazan — Notes on the Identification and Assessment of Hazards,* Institution of Chemical Engineers, 1983, p 55.

8. D Payne et al, *An Index of Electronic Equipment Operability,* American Institute for Research, Pittsburg, PA, Reports Nos AD 607161-5, 1964.

9. A D Swain, Seminar on Human Reliability, UK Atomic Energy Authority, Risley, 1973.

10. T A Kletz, *J of Occupational Accidents,* Vol 1, No 1, 1976, p 95.

11. As ref 6, Appendix III.

CHAPTER 7

SOME ACCIDENTS THAT COULD BE PREVENTED BY BETTER DESIGN

Don't take square-peg humans and try and hammer them into round holes. Reshape the holes as squares. — P Foley (quoted in *Science Dimension,* Vol 16, No 16, 1984, p 13).

So far, in describing accidents due to human error, I have classified them by the type of error. In this chapter I describe some more accidents that could be prevented by better design, in particular, by changing the design so as to remove opportunities for error, though the errors are of various types.

As stated in Chapter 1, safety by design should always be our aim but often there is no reasonably practicable or economic way of improving the design and we have to rely on improvements to the software. We cannot buy our way out of every problem. However, in this chapter I give examples of cases where changes in design are practicable, at least on new plants, often without extra cost.

I do not, of course, wish to imply that accidents are due to the negligence of designers. Just as we try to prevent some accidents by changing the work situation, so we should try to prevent other accidents by changing the design situation, that is, we should try to find ways of changing the design process so as to produce better designs. The changes necessary will be clearer when we have looked at some examples, but the main points that come out are:

— Cover important safety points in standards or codes of practice.

— Make designers aware of the reasons for these safety points by telling them about accidents that have occurred because they were ignored. As with operating staff, discussion is better than writing or lecturing. (See Section 3.3.3.)

— Carry out hazard and operability studies[1] on the designs. As well as the normal hazop on the line diagrams, an earlier hazop on the flowsheet may allow designers to *avoid* hazards by a change in design instead of *controlling* them by adding on protective equipment.[2]

7.1 Isolation of protective equipment

An official report[3] described a boiler explosion which killed two men. The boiler exploded because the water level was lost. The boiler was fitted with two sight glasses, two low level alarms, a low level trip which should have switched on the water feed pump and another low level trip, set at a lower level, which should have isolated the fuel supply. All this protective equipment had been isolated by closing two valves. The report recommended that it should not be possible to isolate all the protective equipment so easily.

We do not know why the valves were closed. Perhaps they had been closed for maintenance and someone forgot to open them. Perhaps they were closed in error. Perhaps the operator was not properly trained. Perhaps he deliberately isolated them to make operation easier, or because he suspected the protective equipment might be out of order. It does not matter. It should not be possible to isolate safety equipment so easily. It is necessary to isolate safety equipment from time to time but each piece of equipment should have its own isolation valves, so that only the minimum number need be isolated, trips and alarms should be isolated only after authorisation in writing by a competent person and this isolation should be signalled in a clear way, for example, by a light on the panel, so that everyone knows that it is isolated.

In addition, although not a design matter, regular checks or audits of protective systems should be carried out to make sure that they are not isolated. Such surveys, in companies where they are not a regular feature, can bring appalling evidence to light.

One audit of 14 photo-electric guards showed that all had been by-passed by modifications to the electronics, modifications which could have been made only by an expert.[4]

7.2   Better information display

The tyres on a company vehicle were inflated to a gauge pressure of 40 psi (2.7 bar) instead of the reccomended 25 psi (1.7 bar). A tyre burst and the vehicle skidded into a hut. Again we do not know why the tyres were over-inflated. It may have been due to ignorance by the man concerned, or possibly he was distracted while inflating them. However, the factory management found a simple way of making such errors less likely: They painted the recommended tyre pressures above the wheels of all their vehicles.

There is another point of interest in this incident. The company concerned operated many factories, but only the one where the incident occurred made the change. The other factories were informed, but decided to take no action — or never got round to it. This is a common failing. When an accident has occurred *to us* we are willing to make a change to try to prevent it happening again, so willing, that we sometimes over-react. When the accident happened elsewhere, the safety adviser has to work much harder to persuade people

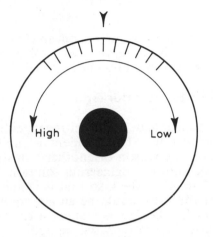

It is better to put the scale on the base-plate instead of the knob.
There is then no doubt which way the knob should be turned.

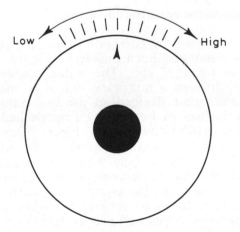

Figure 7.1 Puzzle – which way do you turn the knob to increase the reading?

to make a change. A small fire at our place of work has more effect than a large fire elsewhere in the country or a conflagration overseas.

Another example of prevention of errors by better information display is shown in Figure 7.1.[5]

7.3   Pipe failures

About half the large leaks that occur on oil and chemical plants are due to pipe failures. These have many causes. Here are accounts of a few that could have been prevented by better design.[6]

7.3.1   Remove opportunities for operator errors

The wrong valve was opened and liquid nitrogen entered a mild steel line causing it to disintegrate. This incident is similar to those described in Section 2. For one reason or another a mistake is liable to be made sooner or later and an engineering solution is desirable. The pipe could be made of stainless steel, so that it would not matter if the valve was opened, or the valve could be an automatic one, kept shut by a low temperature trip. Designers would never design a plant so that operation of a valve in error caused equipment to be over-pressured; they would install a relief valve or similar device. They are less willing to prevent equipment getting too cold (or too hot). Design codes and procedures should cover this point.[7]

7.3.2   Remove opportunities for construction errors

A fire at a refinery was caused by corrosion of an oil pipeline just after the point at which water had been injected[8] (Figure 7.2a). A better design is shown in Figure 7.2b. The water is added to the centre of the oil stream through a nozzle so that it is immediately dispersed. However a plant that decided to use this system found that corrosion got worse instead of better. The nozzle had been installed pointing upstream instead of downstream (Figure 7.2c).

Once the nozzle was installed it was impossible to see which way it was pointing. On a better design there would be an external indication of the direction of flow, such as an arrow, or, better still, it would be impossible to assemble the equipment incorrectly. We should design, when possible, so as to remove the opportunity for construction and maintenance errors as well as operating errors.

7.3.3   Design for all foreseeable conditions

A bellows was designed for normal operating conditions. When it was steamed through at a shut-down, it was noticed that one end was 7 inches higher than the other, although it was designed for a maximum

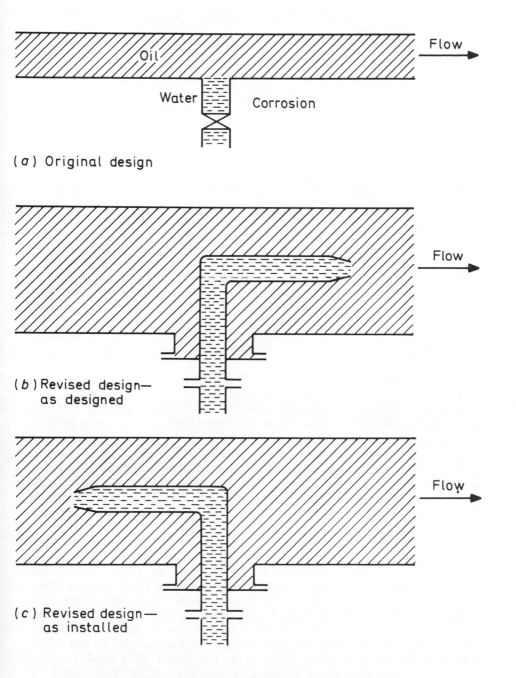

(a) Original design

(b) Revised design—
    as designed

(c) Revised design—
    as installed

Figure 7.2   Methods of adding water to an oil stream.

deflection of ± 3 inches. During normal operation the maximum deflection was 1 inch.

The man who carried out the detailed design of the pipework was dependent on the information he received from other design sections, particularly the process engineering section. The design organisation should ensure that he receives information on transient conditions, such as start-up, shut-down and catalyst regeneration as well as normal operating conditions.

7.4   Vessel failure

These are very rare, much less common than pipework failures; some do occur although only a few of these could be prevented by better design. Many are the result of treating the vessel in ways not foreseen by the designer, for example, overpressuring it by isolation of the relief valve. If designers do fit isolation valves below relief valves, then they are designing an error-prone plant. (Nevertheless, in certain cases, when the consequences of isolation are not serious, isolation of a relief valve may be acceptable.)[7]

Another error made by vessel designers is inadvertantly providing pockets into which water can settle. When the vessel heats up and hot oil comes into contact with the water, it may vaporise with explosive violence. In these cases the action required is better education of the designers, who were probably not aware of the hazard.

7.5   The Sellafield leak

A *cause célèbre* in 1984 was a leak of radioactive material into the sea from the British Nuclear Fuels Limited (BNFL) plant at Sellafield, Cumbria. It was the subject of two official reports[9,10] which agreed that the discharge was due to human error, though it is not entirely clear whether the error was due to a lack of communication between shifts, poor training or wrong judgment. However, both official reports failed to point out that the leak was the result of a simple design error, that would have been detected by a hazard and operability study,[1] if one had been carried out.

As a result of the human error some material which was not suitable for discharge to sea was moved to the sea tanks (Figure 7.3). This should not have mattered as BNFL thought they had 'second chance' design, the ability to pump material back from the sea tanks to the plant. Unfortunately the return route used part of the discharge line to sea. The return line was 2 inches diameter, the sea line was 10 inches diameter, so solids settled out in the sea line where the linear flow rate was low and were later washed out to sea. The design looks as if it might have been the result of a modification. Whether it was or not, it is the sort of design error that would be picked up by a hazard and operability study.

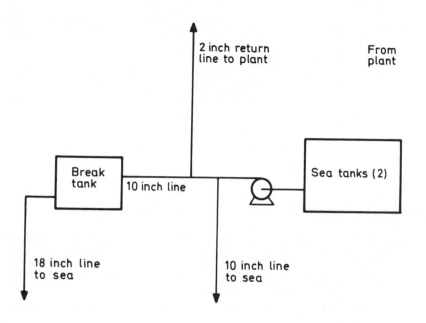

Figure 7.3 Simplified line diagram of the waste disposal system at Sellafield.

The authors of the official reports seem to have made the common mistake of looking for culprits instead of looking for ways of changing the work situation, in this case by improving the design process.

7.6 Domestic accidents

A report on domestic accidents[11] said, "A more caring society would be the ideal but there is no way of obliging people to be more caring". . They therefore gave examples of ways in which better design has reduced or could reduce accidents in or near the home. For example:

- Since 1955 new paraffin heaters have had to be self-extinguishing when tipped over. Since then these heaters have caused fewer fires.
- Many burns are the result of falls onto electric fires. They could be prevented by using fires without exposed heating elements.
- Since 1964 new children's nightdresses have had to have increased resistance to flame spread. Since then fire deaths have fallen;
- Many drownings due to illness or drunkenness could be prevented by providing railings or closing paths at night.

— Many drownings of small children could be prevented by making the edges of ponds shallow, filling in disused reservoirs and canals or closing access through derelict land.

The report concludes that "environmental and product design is the most reliable long-term means of accident prevention . . . Already the design professions seem to take seriously designing for the physically handicapped who are otherwise active. Maybe they will soon see the need to design for the infirm, the mentally ill, drunks, drug addicts, neglected children etc, or indeed anyone who is tired or under some form of stress".

References to Chapter 7

1. T A Kletz, *Hazop and Hazan — Notes on the Identification and Assessment of Hazards,* Institution of Chemical Engineers, 1983.

2. T A Kletz, *Cheaper, Safer Plants or Wealth and Safety at Work,* Institution of Chemical Engineers, 1984.

3. *Boiler Explosions Act 1882 and 1890 — Preliminary Enquiry No 3453,* HMSO, 1969.

4. M Field, *Health and Safety at Work,* Vol 6, No 12, Aug 1984, p 16.

5. A Chapanis, *Man-Machine Engineering,* Tavistock Publications 1965, p 42.

6. T A Kletz, *Plant/Operations Progress,* Vol 3, No 1, Jan 1984, p 19.

7. T A Kletz, *Myths of the Chemical Industry or 44 Things a Chemical Engineer ought not to know,* Institution of Chemical Engineers, 1984.

8. *The Bulletin, The Journal of the Association for Petroleum Acts Administration,* April 1971.

9. *The Contamination of the Beach Incident at BNFL Sellafield,* Health and Safety Executive, 1984.

10. *An Incident Leading to Contamination of the Beaches near to the BNFL Windscale and Calder Works,* Dept of Environment, 1984.

11. B Poyner, *Personal Factors in Domestic Accidents,* Dept of Trade, 1980.

CHAPTER 8

SOME ACCIDENTS THAT COULD BE PREVENTED BY BETTER CONSTRUCTION

How very little, since things were made,
Things have altered in the building trade —
Rudyard Kipling, *A Truthful Song*

In this chapter I describe some accidents which were the result of construction errors; in particular, they resulted from the failure of construction teams to follow the design in detail or to do well, in accordance with good engineering practice, what had been left to their discretion. The reasons for the failures may have been lack of training or instructions or a lack of motivation, but the action needed is the same: Better inspection during and after construction in order to see that the design has been followed and that details left to the discretion of the construction team have been carried out in accordance with good engineering practice.

One construction error — that led to the collapse of the Yarra bridge — has already been discussed (Section 3.3.5).

8.1   Pipe failures

The following examples of construction errors which led to pipe failures (or near-failures) are taken from a larger review of the subject.[1] Many of the failures occurred months or years after the construction errors.

8.1.1   A temporary support, erected to aid construction, was left in position.

8.1.2   A plug, inserted to aid pressure testing, was left in position. It was not shown on any drawing and blew out 20 years later.

8.1.3   Pipes were inadequately supported, vibrated and failed by fatigue. (The operating team could also have prevented the failure, by seeing that the pipe was adequately supported.)

8.1.4   A construction worker cut a hole in a pipe at the wrong place and, discovering his error, patched the pipe and said nothing. The patch was then covered with insulation. The welds, which were not radiographed (though the pipe was supposed to be

100% radiographed), were sub-standard and corroded. There was a leak of phosgene and a man was nearly killed.

8.1.5 A new line had the wrong slope so the contractors cut and welded some of the hangers. They failed. Other hangers failed due to incorrect assembly and absence of lubrication.

8.1.6 Two pipe ends, which were to be welded together, did not fit exactly and were welded with a step between them.

8.1.7 On several occasions bellows have failed because the pipes between which they were installed were not lined up accurately. The men who installed them apparently thought that bellows can be used to take up misalignment in pipes. In fact, when bellows are used, pipes should be aligned more, not less, accurately than usual.

8.1.8 Pipes have been fixed to supports so firmly that they were not free to expand, and tore in trying to do so.

8.1.9 When a pipe expanded, a branch came in contact with a girder, on which the pipe was resting, and was torn off. The pipe was free to expand 5 inches but actually expanded 6 inches.

8.1.10 Insufficient forged T pieces were available for a hot boiler feed water system, so three were improvised by welding together bits of pipe. Four years later they failed.

8.1.11 Dead-ends have been left in pipelines in which water and/or corrosion products have accumulated. The water has frozen, splitting the pipe, or the pipe has corroded.

8.1.12 On many occasions construction teams have used the wrong material of construction. Often the wrong material has been supplied, but the construction team did not carry out adequate checks. The responsibility lies with the construction management rather than the workers.

For example, an ammonia plant was found to have the following faults, none of which were detected during construction and all of which caused plant shut-downs:[2]

1. Turbine blades were made from the wrong grade of steel.

2. The bolts in the coupling between a turbine and a compressor had the wrong dimensions.

3. There was a machining error in another coupling.

4. Mineral-filled thermocouples were filled with the wrong filler.

5. The rivets in air compressor silencers were made from the wrong material.

Other faults detected in time included:

6. Defects in furnace tubes.

7. Blistering caused by welding fins onto furnace tubes.

8. Some furnace tubes made from two different grades of steel.

9. Some furnace tube bends made from the wrong grade of steel.

A list of points to check during and after construction is included in Ref 1.

8.2 Miscellaneous incidents

8.2.1 A compressor house was designed so that the walls would blow off if an explosion occurred inside. The walls were to be made from light-weight panels secured by pop rivets. The construction team decided that this was a poor method of securing the wall panels and used screws instead. When an explosion occurred in the building the pressure rose to a much higher level than intended before the walls blew off, and damage was greater that it should have been.

In this case a construction engineer, not just a construction worker, failed to follow the design. He did not understand, and had probably not been told, the reason for the unusual design.

8.2.2 An explosion occurred in a new storage tank, still under construction. The roof was blown off and landed, by great good fortune, on one of the few pieces of empty ground in the area. No one was hurt.

Without permission from the operating team, and without their knowledge, the construction team had connected up a nitrogen line to the tank. They would not, they said, have connected up a product line but they thought it would be quite safe to connect up the nitrogen line. Although the contractors closed the valve in the nitrogen line (Figure 8.1), it was leaking and a mixture of nitrogen and flammable vapour entered the new storage tank. The vapour mixed with the air in the tank and was ignited by a welder who was completing the inlet piping to the tank.

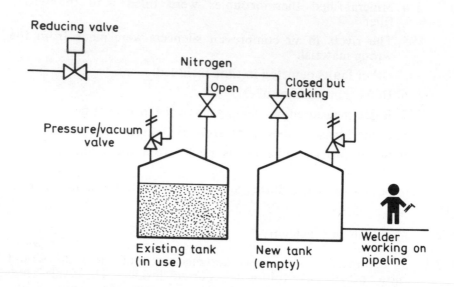

Figure.8.1 Arrangement of nitrogen lines on two tanks. Connection of the new tank to the nitrogen system led to an explosion.

The contractors had failed to understand that:

1. The vapour space of the new tank was designed to be on balance with that of the existing tank, so the nitrogen will always be contaminated with vapour.

2. Nitrogen is a process material, it can cause asphyxiation, and it should be treated with as much care and respect as any other process material.

3. No connection should be made to existing equipment without a permit-to-work, additional to any permit issued to construct new equipment. Once new equipment is connected to existing plant it becomes part of it and should be subject to the full permit-to-work procedure.

8.3   Prevention of construction errors

In the opening paragraph of this chapter I suggested that errors by construction teams are best detected by detailed inspection during and after construction. Who should carry out the inspection? The checks made already by construction inspectors are clearly not sufficient.

The team who will start up and operate the plant have a lot of incentive to inspect the plant thoroughly, as they will suffer the results of any faults not found in time. The designers, though they may not spot their own errors, may see more readily than anyone else when their intentions have not been followed. Perhaps, therefore, inspection of the plant during and after construction should be carried out by the start-up team assisted by one or more members of the design team.

Could we reduce the number of construction errors by taking more trouble to explain to construction workers the nature of the materials to be handled and the consequences of not following the design and good practice? Usually little or no attempt is made to carry out such training and many construction engineers are sceptical of its value, because of the itinerant nature of the workforce involved. However, perhaps it might be tried. Certainly, there is no excuse for not telling construction engineers and supervisors why particular designs have been chosen, and thus avoiding errors such as those described in Sections 8.1.10, 8.2.1 and 8.2.2 above.

Obviously, because of the nature of the task, it is difficult to prevent construction errors by changes in design and the approach must be mainly a software one — better inspection and perhaps training. However we should, when possible, avoid designs which are intolerant of construction errors. Bellows, for example, (Section 8.1.7) may fail if the pipes between which they are placed are not lined up accurately. Fixed piping can be forced into position, but not bellows. Bellows are therefore best avoided, when hazardous materials are handled, and flexibility obtained by incorporating expansion loops in the piping.

Similarly, errors similar to that shown in Figure 7.2 can be avoided by designing the equipment so that it is impossible to assemble it incorrectly.

References to Chapter 8

1. T A Kletz, *Plant/Operations Progress*, Vol 3, No 1, June 1984, p 19.

2. S W Kolff and P R Mertens, *Plant/Operations Progress*, Vol 3, No 2, April 1984, p 117.

CHAPTER 9

SOME ACCIDENTS THAT COULD BE PREVENTED BY BETTER MAINTENANCE

The biggest cause of breakdowns is maintenance — Anon.

In this chapter I describe some accidents which were the result of maintenance errors. Sometimes the maintenance workers were inadequately trained, sometimes they took short cuts, sometimes they made a slip or had a moment's aberration. Sometimes it is difficult to distinguish between these causes, or more than one may have been at work. As with construction errors, it is often difficult or impossible to avoid them by a change in design and we are mainly dependent on training and inspection. However, designs which can be assembled incorrectly should be avoided as should equipment such as bellows, which is intolerant of poor quality maintenance.

9.1 Incidents which occurred because men did not understand how equipment worked

9.1.1 On a number of occasions men have been asked to change a temperature measuring device and have removed the whole thermowell. One such incident, on a fuel oil line, caused a serious refinery fire.[1]

On several occasions, when asked to remove the actuator from a motorised valve, men have undone the wrong bolts and dismantled the valve[2,3]. One fire which started in this way killed 6 men. On other occasions trapped mechanical energy, such as a spring under pressure, has been released.

A hardware solution is possible in these cases. Bolts which can safely be undone when the plant is up to pressure could be painted green; others could be painted red.[4] A similar suggestion is to use bolts with recessed heads and fill the heads with lead if the bolts should not be undone when the plant is up to pressure.

9.1.2 Men have been injured when dismantling diaphragm valves because they did not realise that the valves can contain trapped liquid (Figure 9.1). Here again, a hardware solution is often possible: Liquid will not be trapped if the valves are installed in a vertical section of line.

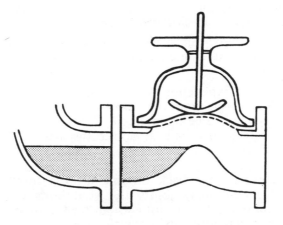

Figure 9.1 Liquid trapped in a diaphragm valve. This can be avoided by locating the valve in a vertical line.

9.2   Incidents which occurred because of poor maintenance practice

9.2.1 It is widely recognised that the correct way to break a bolted joint is to slacken the bolts and then wedge the joint open on the side furthest from the person doing the job. If there is any pressure in the equipment, then the leakage is controlled and can either be allowed to blow off or the joint can be remade.   Nevertheless accidents occur because experienced men undo all the bolts and pull a joint apart.

For example, two men were badly scalded when they were removing the cover from a large valve on a hot water line, although the gauge pressure was only 9 inches of water (0.33 psi or 0.023 bar). They removed all the nuts, attached the cover to a hoist and lifted it.

9.2.2 On many occasions detailed inspections of flameproof electrical equipment have shown that many fittings were faulty, for example, screws were missing or loose, gaps were too large, glasses were broken.

This example illustrates the many-pronged approach necessary to prevent many human error accidents:

Hardware: Flameproof equipment requires careful and frequent maintenance. It is sometimes, particularly on older

101

plants, used when Zone 2 equipment — cheaper and easier to maintain — would be adequate.

Flameproof equipment requires special screws and screwdrivers but spares are not always available.

Training: Many electricians do not understand why flameproof equipment is used and what can happen if it is badly maintained.

Inspection: Experience shows that it is necessary to carry out regular inspections or audits of flameproof equipment if standards are to be maintained. Often equipment at ground level is satisfactory but a ladder discloses a different state of affairs!

9.2.3 When a plant came back on line after a turnaround it was found that on many joints the stud bolts were protruding too far on one side and not enough on the other, so that some nuts were secured by only a few threads.

On eight-bolt joints the bolts were changed one at a time. Four-bolt joints were secured with clamps until the next shutdown.

9.2.4 A young engineer was inspecting the inside of a 60 inch diameter gas main, wearing breathing apparatus supplied from the compressed air mains. He had moved 200 feet from the end when his face mask started to fill with water. He pulled it off, held his breath, and walked quickly to the end.

He dicovered that the air line had been connected to the bottom of the compressed air main instead of the top. As a young engineer, with little experience, he had assumed that the "safety people" and the factory procedures would do all that was required and that he could rely on them.

It is, of course, the responsibility of those who issue the entry permit and those who look after the breathing apparatus to make sure everything is correct, but men should nevertheless be encouraged, before entering a vessel, to carry out their own checks.

A hardware solution is possible in this case. If breathing apparatus is supplied from cylinders or from portable compressors instead of from the factory compressed air supply, contamination with water can be prevented.

Of course, no system is perfect. Those responsible may fail to supply spare cylinders, or may not change them over in time, or may even supply cylinders of the wrong gas. Compressors may fail or be switched off.

We are dependent on procedures, on training to make sure people understand the procedures and on inspections to make sure people follow the procedures. But some designs provide more opportunities for error than others and are therefore best avoided. On the whole piped air is more subject to error than air from cylinders or a compressor.

9.2.5 The temperature controller on a reactor failed and a batch overheated. It was then found that there was a loose terminal in the controller. The terminal was secured by a bolt which screwed into a metal block. The bolt had been replaced by a longer one which bottomed before the terminal was tight.

9.3 Incidents due to gross ignorance or incompetence

As in Chapter 3, some incidents have been due to almost unbelievable ignorance of the hazards.

9.3.1 A shaft had to be fitted into a bearing in a confined space. It was a tight fit so the men on the job decided to cool the shaft and heat the bearings. The shaft was cooled by pouring liquefied petroleum gas onto it while the bearing was heated with a acetylene torch![5]

9.3.2 A fitter was required to remake a flanged joint. The original gasket had been removed and the fitter had to obtain a new one. He selected the wrong type and found it was too big to fit between the bolts. He therefore ground depressions in the outer metal ring of the spiral wound gasket so that it would fit between the bolts (Figure 9.2)! As if this were not enough, he ground only three depressions, so the gasket did not fit centrally between the bolts but was displaced ½ inch to one side.

The fitter's workmanship is inexcusable but the possibility of error in selecting the gasket was high as four different types of joint were used on the plant concerned.

9.4 Incidents which occurred because men took short cuts

Many accidents have occurred because permit-to-work procedures were not followed. Often it is operating teams who are at fault (see Chapter 10) but sometimes the maintenance team are responsible. Common faults are:

— Carrying out a simple job without a permit-to-work.

— Carrying out work beyond that authorised on the permit-to-work.

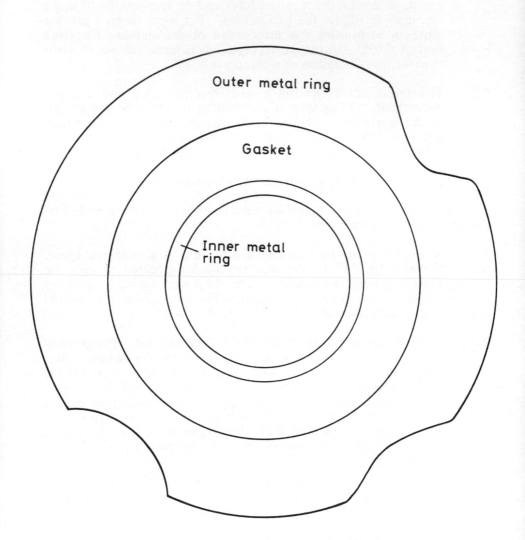

Figure 9.2 A spiral wound gasket was ground away to make it fit between the bolts of a joint.

— Not wearing the correct protective clothing.

To prevent these incidents we should:

— Train people in the reasons for the permit system and the sort of accidents that occur if it is not followed. There is, unfortunately, no shortage of examples and some incidents are described and illustrated in the Institution of Chemical Engineers' Hazard Workshop Module on Preparation for Maintenance.[6]

— Check from time to time that the procedures are being followed, (see Chapter 5).

The following are examples of accidents caused by shortcutting:

9.4.1   A fitter and an apprentice were affected by gas while replacing a relief valve in a refinery. They had not obtained a permit-to-work. If they had, it would have stipulated that breathing apparatus should be worn.[7] The fitter's "excuse" was that he was unaware that the equipment was in use. The fitter and apprentice both smelled gas after removing blanks but took no notice. Many accidents could be avoided if people responded to unusual observations.

9.4.2   Several accidents have occurred because fitters asked for and received a permit to examine the bearings (or some other external part) on a pump or other machine and later decided that they needed to open up the machine. They then did so without getting another permit and a leak occurred. If a permit is issued for work on bearings, the process lines may *not* have been fully isolated and the machine may *not* have been drained.

9.4.3   A permit was issued for a valve to be changed on a line containing corrosive chemicals. It stated that gloves and goggles must be worn. The fitter did not wear them and was splashed with the chemical.

The line had been emptied and isolated by locked valves but some of the corrosive liquid remained in the line.

At first sight this accident seems to be the fault of the injured man and there is little that management can do to prevent it, except to see that rules are enforced. However, examination of the permit book showed that every permit was marked "Gloves and goggles to be worn", though many of them were for jobs carrying no risk. The maintenance workers therefore ignored the instruction and continued to ignore it even on the odd occasion when it was really necessary.

Do not ask for more protective clothing than is necessary. Ask only for what is necessary and then insist that it is worn.

9.4.4 Many accidents have occurred because the operating team failed to isolate correctly equipment which was to be repaired. It is sometimes suggested that the maintenance workers should check the isolations (see Section 9.2.4) and in some companies this is required.

In most companies however the responsibility lies clearly with the operating team. Any check that maintenance workers carry out is a bonus. Nevertheless they should be encouraged to check. It is their lives that are at risk.

On one occasion a large hot oil pump was opened up and found to be full of oil. The ensuing fire killed three men and destroyed the plant. The suction valve on the pump had been left open and the drain valve closed.

The suction valve was chain-operated and afterwards the fitter recalled that earlier in the day, while working on the pump bearings, the chain had got in his way. He picked it up and, without thinking, *hooked it over the projecting spindle of the open suction valve!*

This incident also involved a change of intention. Originally only the bearings were to be worked on. Later the maintenance team decided that they would have to open up the pump. They told the process supervisor but he said that a new permit was not necessary.

9.5 Incidents which could be prevented by more frequent or better maintenance

9.5.1 All the incidents just described could be prevented by better management, that is, by better training, supervision, etc. Other incidents have occurred because of a misjudgment of the level or qualifty of maintenance required. Such incidents are rare in the oil and chemical industries, but occasionally one hears of a case in which routine maintenance or testing has been repeatedly postponed because of pressure of work.

For example, after a storage tank had been sucked in it was found that the flame traps in the three vents were choked with dirt. Although scheduled for 3-monthly cleaning they had not been cleaned for 2 years.

It does not matter if the routine cleaning of flame traps is postponed for a week or a month or even if a whole cycle of cleaning is omitted, but if we continue to put off cleaning an accident in the end is inevitable.

In many cases, if we neglect safety measures, we take a chance and there may be no accident; in the case just described an accident was inevitable.

However a change in design might have prevented the accident. The flame traps had to be unbolted and cleaned by maintenance workers. If they could have been secured without bolts, so that they could have been cleaned by process operators, it is less likely that cleaning would be neglected.

This simple incident illustrates the theme of this book: It is easy to talk of irresponsibility and lack of commitment and to urge engineers to conform to schedules. It is harder, but more effective, to ask why schedules are not followed and to change designs or methods so that they are easier to follow.

9.5.2   A book on railway boiler explosions[8] shows that in the period 1890-1920 the main reasons for these were poor quality maintenance or a deliberate decision to keep defective boilers on the road. Many of the famous railway engineers such as William Stroudley and Edward Fletcher come out badly. Great designers they may have been but — as mentioned in Section 5.2.3 — their performance as maintenance engineers was abysmal; too much was left to the foremen.

Few failures were due to drivers tampering with relief valves or letting the water level in the boiler get too low. Locomotive superintendents, however, encouraged the view that many explosions were due to drivers tampering with relief valves. It is easy to blame the other man.

9.5.3   In 1966 a colliery tip collapsed at Aberfan in South Wales, killing 144 people, most of them children. The official report[9] showed that similar tips had collapsed before, though without serious consequences, and that the action needed to prevent collapse had been well-known for many years.

Following a collapse in 1965 all engineers were asked to make a detailed examination of tips under their control. The Aberfan tip was inspected in the most perfunctory manner. The Inquiry criticised the engineer responsible "for failing to exercise anything like proper care in the manner in which he purported to discharge the duty of inspection laid upon him".

9.6   Can we avoid the need for so much maintenance?

Since maintenance results in so many accidents — not just accidents due to human error but others as well — can we change the work situation by avoiding the need for so much maintenance?

Technically it is certainly feasible. In the nuclear industry, where maintenance is difficult or impossible, equipment is designed to operate without attention throughout its life. In the oil and chemical industries it is usually considered that the high-reliability necessary is too expensive.

Often, however, the sums are never done. When new plants are being designed the aim is to minimise capital cost and often it is no one's job to look at the total cash flow. Capital and revenue may be treated as if they were difficult commodities. While there may be no case for nuclear standards of reliability it may be that a modest increase in reliability would be justified.

References to Chapter 9

1. *Petroleum Review,* October 1981, p 21.

2. T A Kletz, *Hydrocarbon Processing,* Vol 61, No 3, March 1982, p 207.

3. T A Kletz, *The Chemical Engineer,* No 399, Jan 1984, p 33.

4. T A Kletz, *The Chemical Engineer,* No 406, Aug/Sept 1984, p 29.

5. W W Cloe, *Selected Occupational Fatalities related to Fire and/or Explosion in Confined Workspaces as found in reports of OSHA Fatality/Catastrophe Investigation,* Report No OSHA/RP-82/002, US Dept of Labour, April 1982.

6. *Hazard Workshop Module No 004 — Preparation for Maintenance,* Institution of Chemical Engineers.

7. *Petroleum Review,* April 1984, p 37.

8. C H Hewison, *Locomotive Boiler Explosions,* David and Charles, 1983.

9. *Report of the Tribunal appointed to inquire into the Disaster at Aberfan on October 21st, 1966,* HMSO, 1967, paragraph 171.

CHAPTER 10

ACCIDENTS THAT COULD BE PREVENTED BY BETTER METHODS OF OPERATION

"When safe behaviour causes trouble, a change occurs in the unsafe direction" —
K Hakkinen, *Scand J Work Environ Health,*
Vol 9, 1983, p 189.

This chapter describes some incidents which occurred as the result of errors by operating staff. It is not always clear which of the types of error discussed in Chapters 2-5 were involved, nor does it greatly matter. Often more than one was at work.

10.1 Permits-to-work

Accidents which occurred as the result of a failure to operate a permit-to-work system correctly were discussed briefly in Section 5.2.1, while Section 9.4 described incidents which occurred because maintenance workers cut corners.

Here is another incident:

What happened

A permit was issued to connect a nitrogen hose to flange A (Figure 10.1), so that equipment could be leak tested, and then disconnect it when the test was complete.

Flange A was not tagged.

When the leak test was complete the process supervisor (who was a deputy) asked the lead fitter to disconnect the hose. He did not show him the job or check that it was tagged.

What should have happened

Two permits should have been issued: one to connect the hose and a second one — when the time came — to disconnect it.

It should have been tagged. As the job is done several times per year there should have been a permanent tag.

He should have shown him the job and checked that it was tagged.

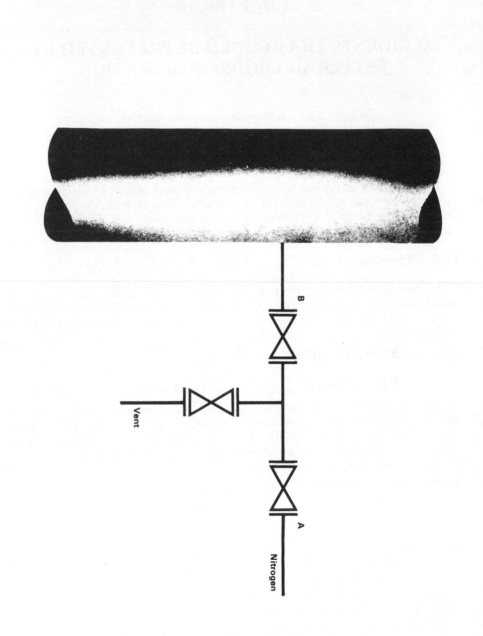

Figure 10.1　　Joint B was broken instead of joint A.

The fitter asked to do the job — not the one who had fitted the hose — misunderstood his instructions and broke joint B. He was new to the Works.	He should have been better trained.
The lead fitter signed off the permit without inspecting the job.	He should have inspected it.
The process supervisor accepted the permit back and started up the plant without inspecting the job.	He should have inspected it.
Toxic gas came out of the open joint B. Fortunately, no one was injured.	

There were at least eight 'human errors':

(1) and (2) Failure of two process supervisors to tag the job.

(3) Failure of the process supervisor to show the job to the second fitter.

(4) Failure of the lead fitter to inspect the completed job.

(5) Failure of the process supervisor to inspect the completed job.

(6) Failure of the manager to see that the deputy process supervisor was better trained.

(7) Failure of the maintenance engineer and foreman to see that the new fitter was better trained.

(8) Failure of the manager and maintenance engineer to monitor the operation of the permit system.

If any one of these errors had not been made, the accident would probably not have occurred. In addition the process supervisor issued one permit instead of two, though issuing two permits would probably not have prevented the accident.

In a sense, (8) includes all the others. If the manager and engineer had kept their eyes open before the incident and checked permits-to-work regularly they could have prevented their subordinates taking short cuts and recognised the need for further training.

Prevention of similar accidents depends on better training and monitoring. However a hardware approach is possible.

If the nitrogen flex was connected by means of a clip-on coupling the maintenance organisation need not be involved at all and the nitrogen supply cannot be disconnected at the wrong point. Alternatively, since the nitrogen is used several times per year, it could be permanently connected by a double block and bleed valve system. The method actually used was expensive as well as providing opportunities for error.

Adoption of the hardware solution does not lessen the need for training and monitoring. Even if we prevent, by a change in design, the particular accident occurring again, failure to follow the permit system correctly will result in other accidents.

10.2 Overfilling tankers

10.2.1 General

Road and rail tankers are often overfilled. If the tankers are filled slowly, the filler is tempted to leave the job for a few minutes. He is away longer than he expected to be and returns to find the tanker overflowing.

If the tankers are filled quickly, by hand, then the operator has only to be distracted for a moment for the tanker to be overfilled. Also, accurate metering is difficult. For these reasons most companies use automatic meters. The quantity to be put into the tanker is set on a meter which closes the filling valve automatically when this quantity has been passed. Unfortunately these meters do not prevent overfilling as:

— the wrong quantity may be set on the meter, either as the result of a slip or because of a misunderstanding of the quantity required.

— there may be a residue in the tanker left over from the previous load which the operator fails to notice.

— there may be a fault in the meter.

For these reasons many companies have installed high level trips in their tankers. They close the filling valve automatically when the tanker is nearly full. Although a separate level measuring device is required in every compartment of every tanker, they are relatively cheap.

10.2.2 Pressure tankers

When liquefied gases are transported, pressure tankers are used. Overfilling does not result in a spillage as the vapour outlet on the top of the tanker is normally connected to a stack or a vapour return line but it can result in the tanker being completely full of liquid. If it warms up, the pressure inside will rise. This does not matter if the tanker is fitted with a relief valve, but in the UK tankers containing toxic gases are not normally fitted with relief valves. On the Continent many tankers carrying liquefied flammabel gases are also not fitted with relief valves[7]. One such tanker was overfilled in Spain in 1978 and burst, killing over 200 people.[1,2] (In addition the tanker is believed to have been weakened by use with ammonia).

Besides overfilling, failure to keep the vapour return line open can also result in the overpressuring of tankers which do not have relief valves. After the insulation fell off a rail tanker while it was being filled, it was found that a valve in the vapour return line was closed and the gauge pressure was 23 bar instead of the usual 10 bar.

Relief valves are not fitted to all pressure tankers as frequent small leaks are considered more hazardous than occasional bursts, particularly on rail tankers that have to pass through long tunnels. However, small leaks can be prevented by fitting bursting discs below relief valves and in the light of the Spanish disaster it seems difficult to justify the continued absence of relief valves. When a cheap, effective hardware solution is available we should not rely on procedures which, for one reason or another, may not always be followed.

10.2.3 A serious incident[3]

Automatic equipment was installed for loading tankers at a large oil storage depot. The tanker drivers set the quantity required on a meter, inserted a card which was their authorisation to withdraw product, and then pressed the start button.

There was a manual valve in each filling line for use when the automatic equipment was out of order. To use the manual valves the automatic valves had first to be opened and this was done by operating a series of switches in the control room. They were inside a locked cupboard and a notice on the door reminded the operators that before operating the switches they should first check that the manual valves were closed.

The automatic equipment broke down and the supervisor decided to change over to manual filling. He asked the drivers to check that the manual valves were shut and then operated the switches to open the automatic valves. Some of the manual valves were not closed; petrol and other oils came out of the filling arms and either over-filled tankers or splashed directly on the ground. The petrol caught fire, killing three men, injuring eleven and destroying the whole row of eighteen filling points.

It is easy to say that the accident was due to the errors of the drivers who did not check that the manual valves were closed or to the error of the supervisor who relied on the drivers instead of checking himself, but the design contained too many opportunities for error. In particular, the filling points were not visible to the person operating the over-ride switches.

In addition, the official report made several other criticisms:

— The employees had not been adequately trained. When training sessions were arranged no one turned up as they could not be spared;

- the drivers had very little understanding of the properties of the materials handled;
- instructions were in no sort of order. Bundles of unsorted documents were handed to the inspector for study;
- there were no regular inspections of the safety equipment.

A combination of error-prone hardware and poor software made an accident inevitable in the long run. To quote from the official report, ". . . had the same imagination and the same zeal been displayed in matters of safety as was applied to sophistication of equipment and efficient utilisation of plant and men, the accident need not have occurred".

10.3   Some incidents that could be prevented by better instructions

This section is concerned with day-to-day instructions rather than the permanent instructions discussed in Section 3.5.   Sometimes care is taken over the permanent instructions but 'anything will do' for the daily instructions.

10.3.1 A day foreman left instructions for the night shift to clean a reactor. He wrote *Agitate with 150 litres nitric acid solution for 4 hours at 80°C.* He did not actually tell them to fill the reactor with water first. He thought this was obvious as the reactor had been cleaned this way in the past.

The night shift did not fill the reactor with water. They added the nitric acid to the empty reactor via the normal filling pump and line which contained some isopropanol. The nitric acid displaced the isopropanol into the reactor, and reacted violently with it, producing nitrous fumes. The reactor, designed for a gauge pressure of 3 bar, burst. If it had not burst, the gauge pressure would have reached 30 bar.

This accident could be said to be due to the failure of the night shift to understand their instructions or use their knowledge of chemistry (if any). It can be prevented only by training people to write clear, unambiguous instructions. No relief system can be designed to cope with unforeseen reactions.

10.3.2 A vented overhead tank was overfilled and some men were asked to clean up the spillage. They were working immediately below the tank which was filled to the top of the vent. A slight change in pressure in one of the lines connected to the tank caused it to overflow again — onto one of the men.

The plant where this occurred paid great attention to safety precautions when issuing permits for maintenance work and would have resented any suggestion that their standards were sloppy, but no one realised that cleaning up spillages should receive as much consideration as conventional maintenance.

10.4   Some incidents involving hoses

Hoses, like bellows (see Section 8.1.7) are items of equipment that are easily damaged or misused, as the following incidents show, and while better training, instructions, inspections, etc, may reduce the number of incidents we should try to change the work situation by using hoses as little as possible, especially for handling hazardous materials.

10.4.1 To extinguish a small fire a man picked up a hose already attached to the plant.   Unfortunately it was connected to a methanol line.

On another occasion a man used a hose attached to a caustic soda line to wash mud off his boots.

In these cases a hardware solution is possible. If hoses must be left attached to process lines, then the unconnected ends should be fitted with self-sealing couplings.

10.4.2 A hose was secured by a screw coupling but only two threads were engaged. When pressure was applied the hose came undone with such force that the end of it hit a man and killed him.

Carelessness on the part of the man who connected the hose? A moment's aberration? Lack of training so that he did not realise the importance of engaging all the threads? It does not matter. Hoses with screw fittings should not be used at pressure.

10.4.3 Hoses often burst or leak for a variety of reasons, usually because the wrong sort was used or it was in poor condition. People are often blamed for using the wrong hose or using a damaged hose. The chance of failure can be reduced by using the hose at as low a pressure as possible. For example, when unloading tankers it is better to use a fixed pump instead of the tanker's pump, as then the hose is exposed to the suction rather than the delivery pressure of the pump.

Why are hoses damaged so often? Perhaps they would suffer less if we provided hangers for them instead of leaving them lying on the ground to be run over by vehicles.

10.4.4 Accidents frequently occur when disconnecting hoses because there is no way of relieving the pressure inside and men are injured by the contents of the hose or by mechanical movement of it.

The hardware solution is obvious. Fit a vent valve to the point on the plant to which the hose is connected. After an incident there is usually a campaign to fit such vents, but after two years hoses are connected to other points on the plant and the accident recurs. It is easy to blame the operators for not pointing out the absence of vent points.

There is a need for a design of vent valve that can be incorporated in a hose as standard. It should not project or it will be knocked off and it should not be possible to leave it open.

10.5   Communication failures

These can of course, affect design, construction and maintenance but seem particularly common in operations, so they are discussed here.

10.5.1 Failures of verbal communication

The incidents described below were really the result of sloppy methods of working, that is, of poor management. People should nor be expected to rely on word of mouth when mishearing or misunderstanding can have serious results. There should be better methods of communication.

For example, a famous accident occurred on the railways in 1873. Two trains were ready to depart from a station. The signalman called out, "Right away, Dick" to the guard of one train. Unknown to him, the guard of the other train was also called Dick.[4]

A flat lorry was backed up against a loading platform and loaded with pallets by a fork lift truck which ran onto the back of the lorry. When the fork lift truck driver had finished he sounded his horn as a signal to the lorry driver to move off. One day the lorry driver heard another horn and drove off just as the fork lift truck was being driven off the lorry; it fell to the ground.

Designers often recommend that equipment is 'checked' or 'inspected' but such words mean little. The designer should say *how often* the equipment should be checked or inspected, *what* should be looked for and *what standard* is acceptable.

After a number of people had been scalded by hot condensate, used for clearing choked lines, it was realised that some operators did not know that 'hot condensate' was boiling water.

116

A member of a project team was asked to order the initial stocks of raw material for a new plant. One of them was TEA. He had previously worked on a plant which TEA meant tri-ethylamine so he ordered some drums of this chemical. Actually tri-ethanolamine was wanted. The plant manager ordered some for further use and the mistake was discovered by an alert storeman who noticed the two similar names on the drums and asked if both chemicals were needed.

A fitter was asked, by his supervisor, to dismantle heat exchanger 347C. The fitter thought the supervisor said 347B and started to dismantle it.

The fitter should, of course, have been shown the permit-to-work, but it is easy to hear a letter incorrectly, particularly on the telephone. It is a good idea to use the international phonetic alphabet shown below. The heat exchanger would have been 347 Bravo, unlikely to be confused with 347 Charlie.

A	ALFA	N	NOVEMBER
B	BRAVO	O	OSCAR
C	CHARLIE	P	PAPA
D	DELTA	Q	QUEBEC
E	ECHO	R	ROMEO
F	FOXTROT	S	SIERRA
G	GOLF	T	TANGO
H	HOTEL	U	UNIFORM
I	INDIA	V	VICTOR
J	JULIET	W	WHISKEY
K	KILO	X	X-RAY
L	LIMA	Y	YANKEE
M	MIKE	Z	ZULU

If there are two pumps on the same duty (one working, one spare) they are often labelled, for example, J25A and J25B. On one plant they were called instead J25 and JA25. Say the names out loud: Jay 25 and Jayay 25 sound too much alike. An electrician asked to replace the fuses in JA25 replaced them in J25 instead.

10.5.2 Failure of written communication

Instructions have been considered in Section 3.5 and 10.3. Figure 10.2 shows three ambiguous notices.

Notice (a) appeared on a road tanker. The manager asked the filler why he had not earthed the tanker before filling it with flammable

This vehicle is fully insulated
Attach no earth

(*a*)

NO ENTRY.
CLEARANCE CERTIFICATE
REQUIRED

(*b*)

There are No Smoking
restrictions

(*c*)

Figure 10.2 Some misleading notices.

liquid. The filler pointed out the notice. It actually referred to the electrical system. Instead of using the chassis as the earth, there was a wired return to the battery. It had nothing to do with the method used for filling the tanker. Notice (b) is error-prone. If the full stop is not noticed, the meaning is reversed.

What does (c) mean?

Pictorial symbols are often used in the hope that they will be understood by people who do not understand English and may not know the meaning of, for example, fragile. However, pictorial symbols can be ambiguous. A storeman saw some boxes marked with broken wineglasses, meaning that the contents were fragile. He took the picture to mean that the contents were already broken and that therefore it did not matter how they were treated. The same man, who lived in a hot country, found some boxes marked with an umbrella, to indicate that they should not be allowed to get wet. He took the picture to mean that the boxes should be kept out of the sun.

10.6   Examples from the railways

10.6.1 Many accidents have occurred because passengers opened train doors before the trains had stopped. For years passengers have been exhorted not to do so. A more effective method of prevention is automatic doors controlled by the crew, used on some suburban electric lines for many years, or doors which are automatically locked when the train is moving. In 1983 British Rail decided that these should be fitted to new Intercity coaches.[5]

10.6.2 While petrol was being offloaded from a train of rail tank wagons, they started to move, the hoses ruptured and 90 tons of petrol were spilt. Fortunately it did not ignite.

The wagons moved because the brakes were not fully applied. They could be applied in two ways:

Automatically: The brakes are then held on by the pressure in compressed air cylinders. As the pressure leaks out of the cylinders the brakes are released. This takes from 15 mins to over 4 hours, depending on the condition of the equipment.

By hand: using a long lever which is held in position by a locking pin.

If the hand brake is applied while the brakes are already held on by compressed air, then when the air pressure falls the brakes are applied more firmly, are hard to release by hand and may have to be dismantled. Instructions therefore state that before the hand brakes are applied the compressed air in the cylinders must be blown off.

However, the operators found an easier way: They applied the hand brakes loosely so that when the air pressure leaked out the brakes would be applied normally.

Unfortunately, on the day of the accident they did not move the hand brake levers far enough.

The design of the braking system was poor. Equipment should be designed so that correct operation is no more difficult than incorrect operation.

The official report[6] did not make this point but it did draw attention to several other unsatisfactory features:

— The siding should be level.

— The hoses should be fitted with breakaway couplings which will seal if the hoses break.

— Emergency isolation valves or non-return valves should be fitted between the hoses and the collecting header.

— The supervisors should have noticed that brakes were not being applied correctly.

References to Chapter 10

1. I Hymes, *The Physiological and Pathological Effects of Thermal Radiation,* Report No SRD R 275, UK Atomic Energy Authority, 1983.

2. H G Stinton, *J of Hazardous Materials,* 1983, Vol 7, p 393.

3. *Official Report on the Fire at the West London Terminal of Esso Petroleum,* HMSO, 1968.

4. A K Steele, *Great Western Broad Gauge Album,* Oxford Publishing Company, 1972.

5. *Modern Railways,* Jan 1985, Vol 42, No 1, p 42.

6. Health and Safety Executive, *Report on a Petroleum Spillage at Micheldover Oil Terminal Hampshire on 2 February 1983,* HMSO, 1984.

7. T.A. Kletz, *Plant/Operations Progress,* Vol. 5, No. 3, July 1986, p 160.

CHAPTER 11

PERSONAL AND MANAGERIAL RESPONSIBILITY

> *Personal responsibility is a noble ideal, a necessary individual aim, but it is no use basing social expectations upon it; they will prove to be illusions* — B Inglis, *Private Conscience — Public Morality*, 1964, p 138.

11.1   Personal responsibility

The reader who has got this far may wonder what has happened to the old-fashioned virtue of personal responsibility. Has that no part to play in safety? Should not people accept some responsibility for their own safety?

We live in a world in which people are less and less willing to accept responsibility for their actions. If a man commits a crime it is not his fault, but the fault of those who brought him up, or those who put him in a position in which he felt compelled to commit the crime. He should not be blamed, but offered sympathy.

This attitude is parodied in the story of the social worker who found a man lying injured by the roadside and said, "Whoever did this to you must be in need of help". And in the story of the schoolboy in trouble who asked his father, "What's to blame, my environment or my heredity?".

Many people react to this attitude by re-asserting that people do have free will and are responsible for their actions. A criminal may say that his crimes were the result of present or past deprivation, but most deprived people do not turn to crime.

For example, one psychologist writes, ". . . criminals aren't victims of upbringing; parents and siblings are victims of the individual's criminality . . . Nor are criminals victims of their peer group . . . young criminals-to-be choose their peers, not the other way around . . . Drugs and alcohol . . . are less the reason why some become criminals than the tool they use . . . to provide themselves the 'courage' to hit a bank . . ."[1]

How do we reconcile these conflicting opinions?

We should distinguish between what we can expect from individuals and what we can expect from people *en masse*.

As individuals we must accept (and teach our children to accept) that we are responsible for our actions, otherwise we are mere computers, programmed by our genes, our parents or society at large.

We must try to work safely, try not to forget, try to learn, do our best.

But as managers, dealing with large numbers of people we must expect them to behave like average men — forgetting a few things, making a few mistakes, taking a few short cuts, even indulging in a little petty crime when the temptation is great,* not to a great extent but doing so to the extent that experience shows people have done in the past. Changing people, if it can be done at all, is a slow business compared with the timescale of plant design and operation. So let us proceed on the assumption that men will behave much as they have done in the past.

11.2   Legal views

This view is supported by the law in the UK, as the following quotations show:

> "(A person) is not, of course, bound to anticipate folly in all its forms, but he is not entitled to put out of consideration the teachings of experience as to the form those follies commonly take".[2]

> "The Factories Act is there not merely to protect the careful, the vigilant and the conscientious workman, but, human nature being what it is, also the careless, the indolent, the weary and even perhaps in some cases the disobedient".[3]

> "The standard which the law requires is that (the employers) should take reasonable care for the safety of their workmen. In order to discharge that duty properly an employer must make allowance for the imperfections of the human nature. When he asks his men to work with dangerous substances he must provide appliances to safeguard them; he must set in force a proper system by which they use the appliances and take the necessary precautions, and he must do his best to see that they adhere to it. He must remember that men doing a routine task are often heedless of their own safety and may become slack about taking precautions.

---

* Reichel, discussing crime in libraries, writes, "Professors who assign research projects that require hundreds of students to use a single source in a library inevitably invite trouble in our competitive and permissive society, unless they make a coordinated effort to provide multiple copies of the source and to subsidize convenient photostat machine operating costs".[15]

He must, therefore, by his foreman, do his best to keep them up to the mark and not tolerate any slackness. He cannot throw all the blame on them if he has not shown a good example himself".[4]

"In Uddin v Associated Portland Cement Manufacturers Limited, a workman in a packing plant went, during working hours to the dust extracting plant — which he had no authority to do — to catch a pigeon flying aroung in the roof. He climbed a vertical steel ladder to a platform where he apparently leant over some machinery and caught his clothing on an unfenced horizontal revolving shaft, as a result of which he lost his arm. The trial judge found that the workman's action was the height of folly, but that the employer had failed to fence the machinery. The judge apportioned 20 per cent of the blame to the employer.

In upholding the award, Lord Pearce, in his judgement in the Court of Appeal, spelt out the social justification for saddling an employer with liability whenever he fails to carry out his statutory obligations. The Factories Act, he said, would be quite unnecessary if all factory owners were to employ only those persons who were never stupid, careless, unreasonable or disobedient or never had moments of clumsiness, forgetfulness or aberration. Humanity was not made up of sweetly reasonable men, hence the necessity for legislation with the benevolent aim of enforcing precautions to prevent avoidable dangers in the interest of those subjected to risk (including those who do not help themselves by taking care not to be injured).

Once the machinery is shown to be dangerous and require fencing, the employer is potentially liable to all who suffer from any failure to fence. And the duty is owed just as much to the crassly stupid as to the slightly negligent employee. It would not be in accord with this piece of social legislation that a certain degree of folly by an employed person should outlaw him from the law's protective umbrella.

The accident in the pigeon case, it is true, would never have happened but for unauthorised and stupid act of the employee. But then the accident would equally not have happened if the machinery had been properly fenced".[5]

The judge added that the workman's actions were not 'actuated by benevolence towards the pigeon'!

The judge's comments suggest that many failures to work safely are deliberate. In fact I think more are due to a moment's forgetfulness. However, the law, like this book, is not primarily concerned with the reasons for failures to work safely but accepts that, for a variety of reasons, men will not always follow the rules and therefore designs and methods of operation should take this into account.

The quotations above were all made during claims for damages by injured employees, that is, under the civil law. Under the criminal law, a manager can be prosecuted if he turns a blind eye to breach of the law, that is, if he sees someone working unsafely and says nothing.

> "Where an offence under any of the relevant statutory provisions committed by a body corporate is proved to have been committed with the consent or connivance of, or to have been attributable to any neglect on the part of, any director, manager, secretary or other similar officer of the body corporate or a person who was purporting to act in any such capacity, he as well as the body corportate shall be guilty of that offence and shall be liable to be proceeded against and punished accordingly".[6]

Figure 11.1 Roof of house in present day Jerusalem showing "battlements".

'Connivance' "connotes a specific mental state not amounting to actual consent to the commission of the offence in question, concomitant with a failure to take any step to prevent the commission thereof".[7]

Finally, it is interesting to note that the law on safety has a long history. The Bible tells us:

> "When thou buildest a new house, then thou shalt make
> a battlement for thy roof, that thou bring not blood
> upon thine house, if any man fall from thence."[8]

In the East the roofs of houses were (and still are) used as extra living space (Figure 11.1).

11.3 Blame in accident investigations

It follows from the arguments of this book that there is little place for blame in accident investigations, even when the accident is due to 'human error'. Everybody makes mistakes or has moments of forgetfulness from time to time and sometimes they result in accidents. Plants should be designed and operated so that these foreseeable errors do not result in accidents (or, if the consequences are not serious, the occasional accident can be accepted). Men should not be blamed for behaving as most people would behave.

Of course, if a man makes repeated errors, more than a normal person would make, or shows that he is incapable of understanding what he is required to do, or is unwilling to do it, then he may have to be moved.

There is also a more pragmatic reason for not emphasising blame in accident investigations. If we do, people will close up (who can blame them?), we shall not find out what happened and we shall be unable to take action to prevent the accident occurring again.

11.4 Managerial wickedness

This book has discussed accidents due to human error and the errors have been of various types, mainly:

Slips and forgetfulness	(Chapter 2)
Ignorance, the result of poor training or instruction	(Chapter 3)
Lack of physical or mental ability	(Chapter 4)
Lack of motivation	(Chapter 5)

In addition, some accidents have occurred as the result of errors of judgment, carelessness, oversights, laziness and inefficiency. The categories of course overlap and often more than one is involved.

I have not included indifference to injury in the list as very few

125

accidents result from a deliberate cold-blooded decision, by managers or workers, to ignore risks to others for the sake of extra profit or output. I say 'cold-blooded' because, in the heat of the moment we are all tempted to take a chance that, if we had time to reflect, we would recognise as unwise, and I say "to others" because we may do to ourselves what we would not do to others.

This view is not universally shared. Many writers seem to believe that the principal cause of accidents is managerial wickedness. For example, some writers (particularly in the safety magazines) look on industrial safety as a conflict between "baddies" (managers) and "goodies" (workers, trade union officials, safety officers, and perhaps factory inspectors). A book intended for trade union members says "Workers have seen how little their lives have often been valued".[9]

This may have been true at one time, but not today. Managers are genuinely distressed when an accident occurs on their plant and would do anything to be able to put the clock back and prevent it. (Dr Who's Tardis would be a useful piece of equipment on every plant). Accidents occur because managers lack knowledge, imagination and drive and are subject to all the other weaknesses that beset human nature but not because they would rather see people hurt than take steps to prevent them getting hurt.

It has been said that the three causes of accidents are:

IGNORANCE,   APATHY   and   AVARICE

In industry, I would agree about the first two but the third is not important.

To conclude this Section, here are two quotations from official reports. The first is from Robens Report.[10]

> "The fact is — and we believe this to be widely recognized — the traditional concepts of the criminal law are not readily applicable to the majority of infringements which arise under this type of legislation. Relatively few offences are clear-cut, few arise from reckless indifference to the possibility of causing injury, few can be laid without qualification at the door of a particular individual. The typical infringement or combination of infringements arises rather through carelessness, oversight, lack of knowledge or means, inadequate supervision, or sheer inefficiency. In such circumstances the process of prosecution and punishment by the criminal courts is largely an irrelevancy. The real need is for a constructive means of ensuring that practical improvements are made and preventative measures adopted."

The second is from the Aberfan Report[11] (see Section 9.5.3):

> ". . . there are no villains in this harrowing story . . . but the Aberfan disaster is a terrifying tale of bungling ineptitude by many men charged with tasks for which they were totally unfitted, of failure to heed clear warnings, and of total lack of direction from above. Not villains, but decent men, led astray by foolishness or ignorance or both in combination, are responsible for what happened at Aberfan."

11.5 Managerial Competence

But if accidents are not due to managerial wickedness, *they can be prevented by better management.* The words in italics sum up this book. All my recommendations call for action by managers. While we would like individual workers to take more care, it is doubtful if they will and therefore we should try to design our plants and methods of working so as to remove or reduce opportunities for error. And if individual workers do take more care, it will be as a result of managerial action — action to make them more aware of the hazards and more knowledgeable about ways to avoid them.

Why then do published accident statistics say that so many accidents — over 50% and sometimes 80 or 90% — are due to 'human failing'.

There are several reasons:

1. Accident reports are written by managers and it is easy to blame the other person.
2. It is easier to tell a man to be careful than to modify the plant or method of working.
3. Accidents *are* due to human failing. This is not untrue, merely unhelpful. We should list only those accident causes we can do something about.
4. There is more scope for improvement in managers than in the managed.

So my counsel for managers is not one of comfort but one of challenge. You can prevent most accidents, not immediately, but in the long run, if you are prepared to make the effort.

Let me end with a few quotations. The first is from a union official:

> "I place the blame for 99 per cent of all accidents fairly and squarely on the shoulders of management, directors, managers, foremen and chargehands, both in the past and right up to the present time.

Unsafe to dangerous practices are carried out which anybody with an observant eye could see if they wished, and if they do see, do they do anything about it? Not until a serious accident happens and then the excuse is "It has never happened before. The job has always been done like this". The workman takes his cue from management. If the management doesn't care the workman doesn't care either until something happens.

Sole responsibility for any accident should be placed fairly and squarely on the shoulders of the Departmental Manager. He should go about with his eyes open instead of sitting in his office, to be able to note unsafe and dangerous practices or work places, and get something done about them as soon as possible. If he gives instructions on these matters he should enforce them."[1] [2]

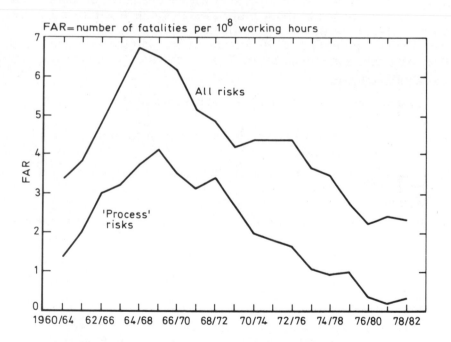

Figure 11.2 ICI's fatal accident rates (The number of fatal accidents in 10^8 working hours or in a group of 1000 men in a working lifetime), expressed as a 5-year moving average, for the period 1960-1982. (From reference 14).

128

The second is from a safety engineer:

". . . outstanding safety performances occur when the plant management does its job well. A low accident rate, like efficient production, is an implicit consequence of managerial control.

Amplification of the desired managerial effect is more certain when managers apply the same vigorous and positive administrative persuasiveness that underlies success in any business function."[13]

Figure 11.2 shows what can be achieved by determined management action. It shows how ICI's fatal accident rate (expressed as a 5-year moving average) has fallen since about 1968 when a series of serious accidents drew attention to the worsening performance.[14]

Finally, a headline from an old copy of the *Farmer's Weekly*:

TAIL BITING IN PIGS

FAULTS IN MANAGEMENT AND SUPERVISION

References to Chapter 11

1. J Page, *Science 84,* Sept 1984, p 84.

2. M Whincup, *The Guardian,* 7 Feb 1966. The quotation is from a House of Lords judgment.

3. A quotation from a judge's summing up, origin unknown.

4. A quotation from a judge's summing up which appeared in several newspapers in October 1968.

5. 'Justinian', *Financial Times,* June 1965.

6. *Health and Safety at Work Act (1974),* Section 37(1).

7. I Fife and E A Machin, *Redgrave's Health and Safety in Factories,* Butterworths, 2nd edition, 1982, p 16.

8. *Deuteronomy,* Chapter 22, verse 8.

9. D Eva and R Oswald, *Health and Safety at Work,* Pan Books, 1981, p 39.

10. *Safety and Health at Work: Report of the Committee 1970-1972 (The Robens Report),* HMSO, 1972, paragraph 26.1.

11. *Report of the Tribunal appointed to inquire into the Disaster at Aberfan on October 21st, 1966,* HMSO, 1967, paragraph 47.

12. F Hynes, *Safety,* (published by British Steel Corporation), August 1971.

13. J V Grimaldi, *Management and Industrial Safety Achievement,* Information Sheet No 13, International Occupational Safety and Health Information Centre (CIS), Geneva, 1966.

14. J L Hawksley, *Proceedings of the CHEMRAWN III World conference (CHEMical Research Applied to World Needs),* The Hague, 25-29 June 1984, Paper 3.V.2.

15. A I Reichel, *J of Academic Librarianship,* Nov 1984, p 219.

CHAPTER 12

THE ADVENTURES OF JOHN DOE

John Doe merely makes the little slips we all make
from time to time. We could tell him to be more
careful. Or we could make simple changes to plant
design or methods of operation, to remove
opportunities for error.

Figure 12.1

Figure 12.2(a)

Figure 12.2(b)

Figure 12.3(a)

Figure 12.3(b)

Figure 12.4(a)

Figure 12.4(b)

Figure 12.5(a)

Figure 12.5(b)

Figure 12.6(a)

Figure 12.6(b)

Figure 12.7(a)

DEFUSED

SLIP-PLATE

Figure 12.7(b)

Figure 12.8(a)

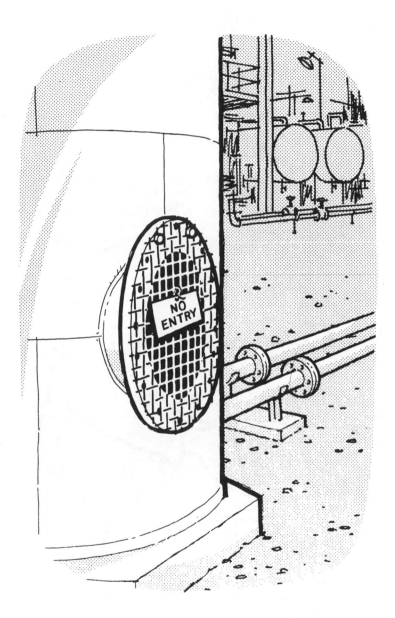

Figure 12.8(b)

Figure 12.9(a)

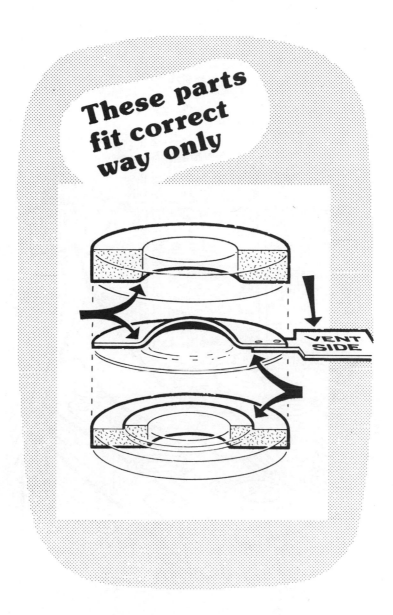

Figure 12.9(b)

Figure 12.10(a)

Figure 12.10(b)

Figure 12.11(a)

Figure 12.11(b)

Figure 12.12(a)

Figure 12.12(b)

Figure 12.13(a)

154

Figure 12.13(b)

POSTSCRIPT

" . . .there is no greater delusion than to suppose that the spirit will work miracles merely because a number of people who fancy themselves spiritual keep on saying it will work them" — L P Jacks, *The Education of the Whole Man,* University of London Press, 1931, Cedric Chivers, 1966, p 77.

Religious and political leaders often ask for a change of heart. Perhaps, like engineers, they should accept people as they find them and try to devise laws, institutions, codes of conduct and so on that will produce a better world without asking for people to change. Perhaps, instead of asking for a change in attitude they should just help people with their problems. (See Section 3.6.3.)

APPENDIX

Influences on morale

In Section 4.3 I suggested that some men may deliberately but unconsciously injure themselves in order to withdraw from a work situation which they find intolerable (or, having accidentally injured themselves, use this as an excuse for withdrawing from work). (See reference 10 of Chapter 4.) Perhaps they do not get on with their fellow workers; perhaps their job does not provide opportunities for growth, achievement, responsibility and recognition. The following notes develop this latter theme a little further, but I want to emphasise that, as stated in Section 4.3, it is not a major contribution to accident rates. It may contribute more to high sickness absence rates.

I want to avoid technical terms as far as possible but there are two that we must use:

Hygienic (or maintenance) *factors* and *motivators.*

Hygienic factors include rates of pay, fringe benefits, working conditions, status symbols and social grouping. *They have to reach a minimum acceptable standard or employees will be dissatisfied and will not work well, but improving these factors beyond the minimum will not make anyone work better.* The minimum acceptable standard varies, from time to time, from place to place and from man to man.

For example, men will not work as well as they might if the pay is low, there is no security, the workplace, canteens and washrooms are filthy and they are treated like dirt. All these factors must be brought up to an acceptable standard before we can hope to get anyone to work well. But improving them alone will not persuade men to work; some companies have given all — high pay, security, pensions, good holidays, fine canteens, sports fields, clubs, childrens outings, the lot but still staff and payroll remain 'browned-off and bloody-minded'.

What's gone wrong?

The answer, according to Herzberg,[1] is that to get men to work well we must:

First, bring the hygienic factors up to scratch (if they are below it).

Second, *fulfil men's need for growth, achievement, responsibility and recognition. Give them a job with a definite aim or object* rather than a collection of isolated tasks, involve them in deciding what that object should be and how it is achieved, give them as much freedom as possible to decide how they achieve that object, show them how their work fits into the wider picture, tell them when they have done a good job and make it clear that there are opportunities for promotion.

This theory explains why we are so well motivated in war-time (the aim or object is obvious) and why process workers, who have the feeling they are driving the bus, are better motivated than maintenance workers. The theory is easier to apply to staff jobs than payroll ones but nevertheless let us try to apply it to the floor-sweeper. The object is obvious and unexciting. What about involvement? It is usual to give the floor-sweeper a schedule, drawn up by the work-study officer, which says which areas are to be swept each day. Why not involve the floor-sweeper in drawing up the schedule? He knows better than anyone which areas need sweeping daily, which weekly. How often does the manager (or anyone else on the plant) thank him for keeping it so clean and make him feel he is "in charge of the plant". (He forgets to add 'for cleaning' when he repeats it to his wife.)

Some men attach more importance to hygienic factors than others. The 'maintenance seeker' is seeking all the time for

More salary

Better working conditions

More "perks"

More status trappings

More security

Less supervision.

He knows all the rules and all the injustices, little and big. If asked about his job, he describes the working conditions, perks etc.

The 'motivation seeker' on the other hand is motivated by the work itself rather than the surroundings. He is seeking all the time to complete more jobs, solve more problems, take more responsibility, earn recognition.

But men are not born maintenance seekers or motivation seekers. Surroundings rich in opportunities for satisfying motivation needs breed motivation seekers and vice versa.

At one time I used to attend regular meetings of works managers at which the lost-time accident rates were displayed. If a works had a high accident rate, the works manager would often explain that there had not really been an increase in accidents but that a number of men had decided to take time off after suffering minor injuries. The production director would then ask, "Why do your men lose time over minor injuries when the men in other works do not?"

Reference

1. F Herzberg, *Work and the Nature of Man,* Staples Press, 1968.

FURTHER READING

A Chapanis, *Man-Machine Engineering,* Tavistock, 1965. The theme is very similar to the theme of this book but the examples are taken from mechanical and control engineering rather than the process industries.

J Reason and K Mycielska, *Absent-Minded? The Psychology of Mental Lapses and Everyday Errors,* Prentice-Hall, 1982. An account of the mechanisms underlying everyday and more serious slips. For a shorter account see J Reason, *Little Slips and Big Disasters,* Interdisciplinary Science Reviews, Vol 9, No 2, 1984, p 179-189.

E Edwards and F P Lees, *Man and Computer in Process Control,* Institution of Chemical Engineers, 1973.

Ergonomics Problems in Process Operation, Symposium Series No 90, Institution of Chemical Engineers, 1984.